Mandy

D1374625

the backyard poultry book

by andrew singer

Published in 1976 by

PRISM PRESS
Stable Court
Chalmington
Dorchester
Dorset DT2 OHB

Copyright 1976 Andrew Singer

Printed by

UNWIN BROTHERS LIMITED
The Gresham Press
Old Woking
Surrey

SBN 0 904727 17 3 Hardback Edition
SBN 0 904727 18 1 Paperback Edition

Illustrations

The illustration on Page 11 is taken from BOOK OF BRITISH
BIRDS, published by Drive Publications Limited, London.
The illustrations on the following pages are taken from WRIGHT'S
BOOK OF POULTRY - 6, 9, 14, 16, 26, 35, 42, 55, 63, 69, 86, 88, 110.
The illustrations on the following pages are taken from 1800
WOODCUTS BY THOMAS BEWICK AND HIS SCHOOL, published
by Dover Publications - Title page, 36, 44, 64, 85, 90, 91, 94, 97,
99, 100, 101.

acknowledgements

Many people have helped with advice and ideas. They all deserve
my thanks, but in particular I would like to mention :

Lady Eve Balfour of Leiston, Suffolk.

J.Bower, Honorary Secretary of the Farm and Food
Society.

J.Darwent, Sales Manager of Hubbard Ireland, Stroud,
Glos. - chick breeders.

Alan Gear of The Henry Doubleday Research Association,
who read the manuscript and made important criticisms.

F.Roscoe of Denby Dale, Yorkshire.

Len Street of St. Blazey, Cornwall, my co-author on The
Backyard Dairy Book, and who helped particularly on finding illustrations.

M.A.Thompson of Shipton, Dorset: a small independent
chick supplier.

Robert and Brenda Vale of Witcham, Cambs.

I would also like to acknowledge the help of various officials of
the Ministry of Agriculture, London, and of their library staff.
Colin Browning was particularly keen that his illustrations should
be both informative and attractive, a task in which I think he has
succeeded. My publisher, Colin Spooner, has been more than usually helpful and understanding. He deserves the credit for having
encouraged me to produce a second Backyard book.

Lastly, I must thank my wife, to whom this book is dedicated.
She has more natural stockmanship than I shall ever have. She
curbed my more ridiculous crackpot ideas that would otherwise
have led the reader astray, (though you may notice that one or
two did get past her) and she also helped in choosing illustrations.

about this book

This is the second of Prism Press's 'Backyard' series of practical books on self-reliance. The first was 'The Backyard Dairy Book' which Len Street and I wrote in 1972. The approach in both books is pratical and enthusiastic, based both on experience and on thorough research into the available literature.

In the dairy book, however, we attempted only an introduction to complex subjects of cow and goat husbandry, butter making, cheese making etc., the aim being to enthuse the reader and to provide enough information to get him started, leading him to more detailed reference books on these subjects for later.

Small scale poultry-keeping is not such a complex subject. The aim of this present book is to provide a complete book of instruction and reference for the backyarder. Nevertheless, plenty of references are given on specialised related subjects.

Having said that, there are bound to be omissions and information that can be improved or updated. In the three years since the dairy book first came out, many readers have sent suggestions for change and improvement, most of which we incorporated into the recently published second edition. I would like to be able to do the same with the poultry book. If you have any suggestions, please send them to me via Prism Press.

I should, right at the start, explain the coverage of this book. The emphasis is on keeping poultry for eggs rather than for meat. I treat meat as a by-product and deal with it in Chapter Nine. The emphasis is also strongly on the domestic fowl, the hen. Ducks, geese, turkeys and other poultry are dealt with in Chapters Eleven to Fourteen, but I assume that the reader starts off with hens.

The book is divided into subject chapters in the normal way, but the chapters have been organised into four parts. There is certain information you ought to have even before you decide definitely to start poultry-keeping. This information is in 'First Stage: Making a Start'. Then there is information which you need once the chicks actually arrive, this is in 'Second Stage: Your Chicks Arrive'....and so on.

There is a great deal of detailed information in this book and I am conscious of the difficulty of remembering all the relevant bits at crucial times. That is why you will find two check lists at the back. One to check that everything is ready before the chicks arrive, and the other listing all the regular routines of poultry-keeping just in case you get lax.

Now read on. I hope you enjoy the book, and more to the point, that the book helps you enjoy your poultry-keeping.

Andrew Singer,
Coleshill, Oxon.
October 1975.

contents

FIRST STAGE: MAKING A START

first stage: making a start

1 why keep poultry?

There are several good reasons:
* They save you money on eggs, poultry-meat, garden fertilizer, and maybe feather cushions as well.
* The eggs and meat you produce taste better, (assuming that they are kept properly, and we will come to that later), and are possibly better for you.
* It avoids relying on what many people consider a repulsive and inhumane system of husbandry, namely laying batteries and broiler houses.
* Keeping poultry is an enjoyable and satisfying hobby.

What about reasons for not keeping poultry?
* Your local by-laws may not allow it.
* Keeping poultry involves work.
* Poultry cannot be left, they need regular attention.
* It is a risk - they may all die young and you lose money.
* Some medical opinion suggests that eggs are not all that good for you anyway.

Here are the details:

THE CASH SAVING

Based on the latest Government produced Family Expenditure Survey, Mr. and Mrs. Average British Household spend, at

1975 prices, an average of around 42p per week on eggs, and £1-31 per week on poultry-meat. When you keep poultry, you tend to eat more eggs and poultry-meat - that's one of the pleasures of it - so the potential savings tend to be more. People who keep poultry tend also to be gardeners, and thus the poultry manure comes in very useful by saving them money in terms of bought fertilizer or increased garden production. Chicken feathers kept and used for cushions, continental quilts and the like do not save you much on a per-week basis, but they are a nice extra bonus.

Of course, how much you actually save by keeping poultry depends on how much you have to pay out to keep the poultry. The main element of cost is feeding your poultry, but there are ways and means of keeping feed costs down pretty low, (we go into details in Chapter Five). We reckon our eggs cost us less than 10p per dozen, and at least half of them are large or medium. This compares with 40p in the shops for similar 'free range' eggs.

It has to be admitted that commercial egg production in Britain is one of the most economically efficient sectors of agriculture. Although you can clearly save money by substituting your own labour and capital for those of the egg-farmer and by using more free food, the cash saving over the year is unlikely to be over £50 - £70 for an average family. To me personally, the cash saving is the least important of the four reasons I give here for keeping poultry.

THE IMPROVED TASTE AND NUTRITIONAL VALUE

This is a personal one, of course. There is no question in my mind that our hens' eggs are better tasting than any bought eggs. The yolks are darker and much richer in flavour, particularly in the summer. The consistency and taste of a really fresh egg white, poached or boiled, is also greatly superior to a shop bought equivalent.

Because of the relative strengths of the commercial interest groups involved, there has been very little scientific work done on the differences between free range and battery eggs. Recent Ministry of Agriculture work suggests that battery eggs have 70 per cent less Vitamin B12 and 50 per cent less Folic Acid than free range, both nutrients essential to red blood cell formation. Note that Vitamin B12 is mainly found in food of animal origin, so perhaps vegetarians should

2

be particularly keen on keeping their layers on free range.

In spite of my scientific training, I have to admit that I do not really need experimental confirmation of what my taste buds tell me. My common sense tells me too. We are what we eat - and the hen is no different. Nature has adapted her to a diet in the wild of live insects, whole grains, seeds and leaves gathered in great variety on the forest floor. She also uses sunlight. Her egg laying system has similarly been adapted to get maximum nutritional value into the egg from this diet, so as to give the developing chick embryo the best chance of survival. That's natural selection, after all. My common sense tells me that a hen fed, in sunlight, on the nearest possible diet to the natural one must produce better eggs, in every way, compared to the hen kept out of sunlight and fed on a single, boring, compound mash consisting mainly of ready crushed grains with various manufactured additives.

As to the taste of free range poultry meat, most readers, particularly the younger ones, will be in for a completely new experience. Virtually the only poultry meat now available in shops is produced in broiler houses. They are killed at eight weeks, by which time they have a live weight of around 4lb. It is claimed that the feed-to-meat ratio is just over 2:1. In fact, a dry feed-to-dry meat ratio would be closer to 16:1 - the difference lying in the carcase weight and the water in the flesh. So, what you are eating is only inefficiently converted compound feed. Compare the taste and consistency of a broiler chicken with any wild game bird: partridge, pheasant, pigeon, or wild duck, and you will understand that there is a world of difference.

It is all a question of the relative value you place on 'tenderness' and taste. The broiler is produced so as to be as tender as a lump of gelatine - but of course it has about as much taste, too. You can roast or even grill it of course, but what have you gained? I prefer the taste of a two year old hen who has lived out her useful life as a layer. She will not roast, but what a fine casserole she makes. The meat actually has its own flavour independent of the sauce.

AVOIDING POULTRY AGRIBUSINESS

"To confine, whom nature has given the urge to scrap, to perch, to flap her wings, to take dust baths, in a wire cage in which she cannot do any of these things, is revoltingly cruel, and I cannot bring myself to talk to anybody who does it, nor would I,

3

on any condition, allow such a person inside my house."
John Seymour in 'Self-Sufficiency'.

Well, that is how upset some people get about battery cages.
As a small diversion from the main theme of this book, I think
it will be instructive to explain why.

No branch of agriculture has been as dramatically and widely
affected by the concepts of agribusiness as poultry-keeping. Under the old traditional farming concept, poultry was considered
as just one small element of the balanced mixed farm. Hens
grazed and harrowed the pastures after other livestock, con-
sumed the dirty wheat and weed-seed mixture that went through
the sieve. On this they provided a useful year-round farm income.

After the First World War, many ex-officers used their gra-
tuities to start specialist poultry farms, in those days mainly
free range. That continued into the 1950's, with a gradual switch
to housed systems, mainly 'deep litter', where the hens led a
reasonably free existence on a large straw or peat-moss floor,
but deprived of sunlight and wild foods. The big change came in
the 1960's and early 1970's. The battery system swept the board.

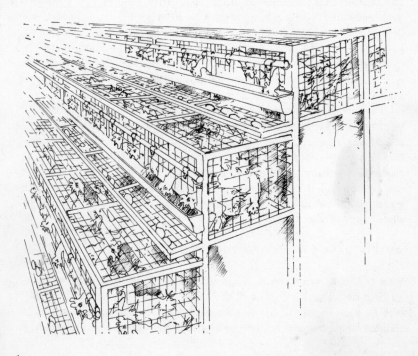

In 1960-61, 19.3 per cent of our eggs were produced in batteries. By 1973-74, the figure had risen to 89.8 per cent, and it is still rising. At the same time, the small independent producers have been declining in importance, whilst the big poultry agribusiness corporations have taken an increasing share of the egg market. Four packing organisations now handle 70 per cent of all packed eggs. 60 per cent of the national laying stock is kept on 2000 holdings, all with a capacity of at least 100,000 birds.

The picture shows a typically commercially available battery system. The average stocking rate has been steadily increasing in the interests of improved return on capital investment. The original idea was for each cage to hold one bird. The average now is about three hens to each of these 16 inch by 18 inch cages. If that is difficult for you to stomach, let me emphasis that this is the average. Some producers must cram more than three in each cage, dreadful though it may seem.

Battery hens have no need of their beaks, of course, since they feed exclusively on mash. They get so bored and frustrated in those frightful prison cages that they have a tendency to peck at each other as substitutes for the insects and grains they miss. The agribusiness answer to this unfortunate and costly habit is, logically enough, to de-beak them.

An official committee reporting on intensive systems in 1965 stated that they were convinced that de-beaking caused considerable pain, lasting much longer than the second or so that the operation takes to perform with a special hot knife. But I doubt that this has stopped a single battery producer from doing it. It must be clear to the average humane open minded reader that once a poultry keeper accepts the generally current attitude that cramming hens in battery cages is wholly justifiable because it produces cheaper eggs, then there is no brake on him resorting to just about any means to get eggs out even more cheaply. He inevitably becomes desensitised to concepts of 'cruelty to hens' - he has to. De-beaking, in my view, is an example of this. Another is witholding water from hens to bring on their moult. Another is increasing the temperature inside the battery house to reduce food consumption. All these practices are widespread.

Perhaps the most extreme example of the agribusiness approach, which would be funny if it was not so inhumane, is the experimental programme at the University of Connecticut to rear featherless chickens for the broiler industry. Fortunately it has failed - for various reasons, not least being that the featherless cockerels are unable to balance adequately on the hens while mating.

Old Sussex Cramming Machine

Cruelty to poultry is not new. The old 'Surrey Chickens' were produced by force-feeding food into them through a tube. Turkeys used to be plucked alive because they believed that it made the meat more tender. But never before, I contend, has cruelty to poultry been so widespread, and so tolerated by public opinion.

AN ENJOYABLE AND SATISFYING HOBBY

Chickens are really not very intelligent creatures. But they have attractive ways. They cluck contentedly when they lay you eggs and expectantly when they see you coming to feed them. Rearing new chicks in the Spring is a special joy, particularly watching the broody take them on scratching forays, teaching them to look for food, and clucking at them to warn them against straying too far. But perhaps the most satisfying experience of all is collecting the day's new-laid warm eggs from the straw nesting boxes. Eggs are definitely one of nature's finest aesthetic works: in shape, texture and colour.

Poultry-keeping is a productive hobby, where the success is easily measured by how many eggs you get, and highly dependent on your personal capability as a poultry keeper. Doing it successfully, getting plenty of eggs and few problems, is very satisfying.

. . . . . AND THE DISADVANTAGES

Now for a little more on the snags. Firstly, you cannot keep poultry just anywhere. In many areas, particularly in towns, Local Authority regulations prohibit it, so that is the first thing that you had better check on. Next, it does take up space. How much depends on what system you use, (see Chapter Three), but you can keep six hens in as small an area as 6 feet by 4 feet, though much more land is preferable if you want really tasty 'free range' eggs.

Keeping poultry does involve work. The main work is in setting it up: building the hen house, fencing them in and finding sources of feed. Apart from that it takes about ten minutes per day to feed and water them, and to collect the eggs. Then ten minutes every week to clean out the droppings and renew the nesting material. I do not consider that work. It is a pleasant distraction from the longer, more demanding tasks of the day. (I have been out three times to feed my hens, collect their eggs or just to observe them since I started writing this chapter.)

Keeping poultry, like most animals, does generally tie you to a regular routine of feeding, watering and egg collecting. But if you are like us and go away for weekends pretty regularly, there are ways in which you can get round this problem and be able to leave them for up to a week without attention, (see Chapter Three). Normally, most people only go away for longer than that once or twice a year, on holidays. Then you will need to persuade a neighbour to do the chores for you. We have found that most consider this a very minor chore, bearing in mind the attraction of free eggs for a fortnight.

Poultry are a risk, of course. You may spend money on buying chicks, feed them for six months, only to have them killed by a fox before they come into lay. It does happen, but not too often. We have suggestions for fox-proofing in Chapter

Six. If you do lose them, the main losses are your time and
the psychological effect. But even at six months, you will only
have lost less than £1 per bird in cash costs. Start again, and
it should not be too long before you recoup the cash loss in eggs
laid. If the fox gets them after about eight months of age, you
have not really lost anything but potential profit.

All these, then, are pretty minimal snags considering the
advantages of keeping poultry. In my view, the only serious
argument against it is that keeping poultry encourages increased egg consumption, which may be bad for your health. I
stress 'may'.

EGG CONSUMPTION AND HEALTH

Certain medical studies attempting to find possible causes for
the high present-day incidence of coronary heart disease, particularly in the developed countries, have suggested a possible
link with the level of dietry intake of 'cholesterol', mainly via
eggs, dairy products and animal fats. These studies have rec-
ieved wide publicity, particularly in the U.S.A., where a survey recently showed that, as a result, one in four adult Amer-
icans believe that it has been proven or possibly proven that
eating eggs is a cause of heart disease. But is this belief in any
way justified?

In this present state of confusion, even amongst experts,
"you pays your money, and you takes your choice". Perhaps you
may feel that even the suspicion of a link between high egg consumption and heart disease warrants keeping egg consumption
low. Well, for what it is worth, I doubt that this is wise.

After all, eggs are a natural food, (particularly if produced
'free range'), nutritionally second only to mother's milk. They
have been part of our regular diet now for many generations. In
my own mind, I am pretty sure that the sudden rise in coronary
heart disease is more likely to be connected with equally dramatic changes in our ways of eating and living, than with egg eating; for instance with sedentary habits, high stress, vastly increased consumption of sugar and refined foods, and the like.
But I cannot prove it to you, of course. I can only say that I have
no hesitation in regularly eating a dozen of our hens' eggs per
week, so long as these are part of a balanced diet.

So what about all these advantages and disadvantages taken
together. If you don't mind the bit of work involved, or the tie
of keeping animals, backyard poultry-keeping is the next logical

step in self-sufficiency beyond vegetable and fruit growing, with which, by the way, it combines excellently, (see Chapter Three on Systems). Even if you are not a self-sufficiency enthusiast, backyard poultry-keeping is a very satisfying hobby which saves you money (unlike most hobbies) and improves the quality of the food you eat. How can you resist?

2 a little technical background

Before we get into the details of feeding, housing and the like, it is as well that you have a basic understanding of the natural systems we are trying to plug into by keeping hens.

THE LIFE-PATTERN OF THE FOWL

The domestic hen, Gallus Domesticus, is the most numerous species of bird on Earth; between eight and nine billion of them are born every year. It is descended from the jungle fowl of Asia, Gallus Gallus, which still exists in the wild.

The wild jungle fowl evolved naturally for a life in the Asian tropical forests. Its large feet and sharp nails enable it to scratch up the surface material of the forest floor to search for a variety of small crawling insects, worms etc. and for plant seeds. These it picks up using a sharp beak and a very fast, hard pecking action of the neck. It also eats tender young shoots found on the forest floor. Operating in flocks, it has a well-developed social sense. Each flock is headed by a dominant male and has a definite 'pecking order'. As with most 'dominant male - many females' species, the male bird is physically much better developed and very protective of the flock. He also has all the worst 'male chauvinist' characteristics - he is aggressive, egocentric, randy, and of course totally uneconomic.

The habitat of the wild jungle fowl subjects it to many predators. Its defences against them are to stay in flocks, its fine hearing, and its wings. These wings are not very much use for

THE ANCESTOR OF THEM ALL

Centuries of domestication have produced 70 different breeds of chicken from the wild red jungle fowl (top). These include: (1) the old English game-bird; (2) the black Minorca; (3) the Sebright bantam; (4) the Rhode Island red; (5) the white Wyandotte; and (6) the game-bird bantam.

flying, but they do give the bird a turn of speed on the run and enable it to get off the forest floor onto the branches of trees. For safety at night, it sleeps perched on branches (called roosting).

Most of its day is taken up with searching for its tiny foods, of which it naturally requires quite a quantity, but it does stop from time to time to indulge in preening and dust bathing. Preening means stroking the feathers with the beak to clean them and oil them. The oil is picked up by the beak from the preen gland, situated near the back of the tail. Dust bathing consists of finding or scratching out a nice dust hole in the ground, squatting into it and fluffing dust into the feathers. The purpose of dust bathing, which is not done by many other types of bird, is not fully understood, but may be connected with cleaning the smaller feathers and keeping them free of parasites. Both preening and dust bathing are obviously forms of relaxation, which tend to happen when the birds are feeling full or sleepy. They are very quiet and social activities.

The digestive system of the fowl is ideally adapted to its diet and habits. It is not designed to cope with bulk foods. The insects, seeds etc. are not digested immediately. They are stored in a sac under the neck called the crop. The complex processes necessary to digest a mixture of various fibrous and horny materials; flesh, greenstuff, seed germ etc., takes place partly in the stomach, where they are chemically broken down and partly in a special object called the gizzard, where they are physically crushed. The gizzard is a sac containing insoluble grit taken down with the food, and enclosed by very strong muscles. In here, the hard bits are crushed and ground by these muscles, using the grit as a grinding medium. A sort of internal mill - we will come to the significance of that later.

It is the reproductive system of the fowl that we plug into, of course, in egg production. Reproduction outside the body is a more primitive system than reproduction inside the body, as takes place in mammals. It has been retained in birds probably because it reduces the weight load in the mother, thus improving her mobility for foraging and avoiding predators. The egg of the jungle fowl is laid in small hollowed-out corners of the forest floor, probably in a really dark corner, near a tree trunk well out of sight of predators. A clutch of about six eggs are laid, one every day or so, and then the hen 'goes broody'. That means that she sits on the eggs, keeping them nice and warm, thus encouraging the growth of the chick embryo inside. She stays that way for twenty-one days, after which the chicks hatch out, already fully covered with down, and with enough food inside them from the egg to last a day or two. By that time, they can walk and pick up the food that the mother hen scratches for them and breaks into smaller pieces if necessary.

That, then, is the life pattern of the jungle fowl. Such birds were first kept domestically well over four thousand years ago, and since then the basic jungle fowl has been developed into an enormous variety of strains of domestic fowl. In many parts of the world, interest has centred mainly on the fighting exploits of the male, with eggs and meat very much secondary attractions. Only in the last hundred and fifty years have really effective specialised egg-laying and meat-producing strains been developed. So these developments are pretty recent compared to the long periods over which the jungle fowl evolved.

In my view, the modern domestic fowl is very much the same bird as the jungle fowl, so far as its life pattern and eating habits are concerned. It still has the same desire to scratch, peck, preen itself, dust bath and roost. I maintain that our systems of keeping fowl should allow for these desires. And frankly, I would hate to see any of them bred out - perfectly possible, of course, with modern genetic techniques and perhaps advantageous for battery producers - but not desirable.

If you kept a jungle fowl in captivity and kept it laying by taking away the eggs as laid, it would lay about thirty eggs per year, going broody at least twice in the process. Peasant farmers in less developed countries still keep fowl very little different from jungle fowl that give them thirty to fifty eggs per year on a diet of what they can scratch in the farmyard. By careful breeding, mainly in Italy early on, but also in the U.K. and America after 1850, the modern fowl's production rate has been dramatically increased. The U.K. national average in 1973-74 was 228 eggs per year. The World record is 361 eggs in 365 days.

Apart from its capacity to lay more eggs, the only other important differences that the domestic laying fowl displays from the wild jungle fowl are its adaptation to colder climates, and its capacity to take in more food to produce the eggs. However, its digestive system is still very much the same. It is suited to whole grains, which are, after all, just big seeds, but it is not a bulk-feeding animal, like a cow or a duck.

Although the modern hen can be persuaded to lay a lot more eggs, it still is not quite an all-year round once-a-day machine process.

The egg yolk develops in the ovary of the fowl and when ready, is taken down by the mouth of the oviduct, where the white and then the shell are added gradually. The egg emerges from the vent, the large opening below the bird's tail, which is also where the faeces emerge. The passage of the egg down the oviduct takes twenty-two hours and in a regularly laying bird, an egg is laid on average every twenty-six and a half hours, though this varies from twenty-five and a half in Spring to twenty-seven and a half in Winter.

A modern hen kept inside in reasonably natural conditions will come into lay when the days start to lengthen in early Spring. In January and February she will lay at about forty to fifty per cent (taking one hundred per cent as one egg per day), rising fast to eighty per cent in April and May. After this Spring peak, she gradually reduces to between sixty and seventy per cent by July or August, and then around September she goes 'into moult'.

Ovary and Oviduct

The days becoming shorter tell her that the laying season should be over; she stops laying and the extra nourishment available to her body allows her to grow more feathers ready for the Winter. The old feathers are pushed out and for a while she looks pretty scraggy. After about six to ten weeks in moult, she may come in to lay again in October/November or she may stay out of lay until January or February.

The important points to notice about this annual pattern are firstly that the moult does not cause the loss of egg production, it is more the other way around. Secondly, egg production is stimulated by the pattern of daylight. Lengthening daylight hours bring the hen into lay; shortening hours take her out of lay. Note that it is the pattern, not the actual amount of daylight hours, that matters.

The modern hen has had her tendency to go broody reduced

by selective breeding. This is an advantage in that egg production is interrupted less when she does go broody - at least for three or four days until you persuade her out of broodiness. It is also, as we shall see later in Chapter Seven, something of a disadvantage to the backyarder who needs broody hens to rear replacement chicks.

The heat of Summer weather and the sight of other hens sitting are two factors known to bring on broodiness. To persuade a hen out of broodiness, you must keep her on a wire-mesh floor with plenty of draughts under it. She eventually realises that no chicks are going to appear in those conditions.

Although the egg-laying pattern of a modern hen kept under natural conditions is determined by the annual cycle of the seasons and by her tendency to broodiness and otherwise, it is also determined by her age. A hen reaches sexual maturity at about 21 weeks, when she is capable of laying at about ten per cent of her eventual capability. She does not reach peak capability until 28 weeks.

If a hen reaches the age of sexual maturity when the days are lengthening, she will come into lay. If the days are shortening, she may not actually start laying until they lengthen again the following Spring. However, this is not universally so.

WHAT IS AN EGG ?

Half of the yolk of the egg consists of fats and proteins, providing the main sustenance for the growing chick embryo. The white is 90 per cent water, but with an extra store of protein. Eggs are very nutritious - they are designed to be. Overall, they are around 14 per cent protein and 10 per cent fat. They also contain iron, calcium and vitamins A, B, D, and E.

Hen eggs are generally either white shelled or brown shelled. There is no difference in their food value. In parts of South America, they have hens that lay blue shelled eggs. If any jet-setting reader were able to get me a few fertile blue hen eggs from those parts, I would be most appreciative.

The reasons for the continuing bias in favour of brown eggs in some parts of the World, for instance the U.K. and New England, and in favour of white ones in others, as in the rest of the U.S.A., are historical. In the U.K., people associate brown eggs with the heavier breeds traditionally kept by the backyarders and free range producers - hence with fresh local eggs. They associate white eggs with the lighter breeds that

first went into batteries. The eggs had little lion marks stamped
on them, and hence were less local, and often less fresh. Gradually, perhaps, they will realise that nearly all brown eggs are
now produced in the same battery houses as are white eggs.
The reason the white ones are cheaper is that the small white
egg-laying breeds are slightly more 'efficient' battery producers,
although the gap is narrowing.

Although wild birds only lay eggs if fertilised by a male, the
domestic hen lays steadily whether mated or not. So you do not
need to keep a cock if all you want is eggs - you only need one
for breeding. Some people claim a difference in taste between
fertilised and unfertilised eggs. I have never noticed it. The
only difference is the tiny white speck on the edge of the yolk
which is the very early beginnings of the embryo.

3 which system to use

In the last Chapter, I went to some lengths to describe the natural life pattern of the fowl. I also made clear in Chapter One my view that any system of keeping poultry in captivity should, so far as possible, allow for the birds' natural habits rather than frustrate them. What this means in practice is that any system should provide:
* plenty of room to scratch about searching for food - preferably including live insects, seeds and young green shoots;
* a nice sheltered, shady spot for communal preening and a dusty spot for dust baths;
* branches or reasonable substitutes for perching and roosting at night;
* a dark dry corner with a soft forest-floor-like surface for laying eggs.

Although the domestic fowl as bred so far is well adapted to our colder climate, some protection from the worst extremes of a British Winter makes for better egg production. After all, their body temperature is 41 degrees C, 106 degrees F, fever level by human standards. The less food they need to maintain this in Winter, the better.

So that, basically, is what they need. Let us now look at how people have, so far, tended to provide these needs.

FULL FREE RANGE

That means giving the hens as much grassland as they need, consistent with not reducing its quality by scratching and the hens' preference for young shoots. They make their own dust baths and preening areas, of course, but are usually provided with a moveable house, containing nest boxes and roosts. The standard usually quoted is 100 birds per acre, or 425 square feet per bird, which means that this system is only suitable if you have got plenty of grassland that you are not too bothered about using very efficiently. But its great advantage is that the birds find lots of live insects, little seeds and shoots, which make for the much-prized 'free range egg'. They also need less bought food, at least in Summer.

FOLDED FREE RANGE

Folding means allowing the hens on part of the grassland at a time, allowing other parts to 'rest' in turn. This means that more new shoots get a chance to grow, and the land does not get too soured by excess quantities of fresh poultry manure. After all, poultry manure is very rich in nitrogen and also pretty acid. Too much of it is a bad thing. Resting the land also kills off poultry parasites and keeps the birds healthier. The usual method of folding consists of confining the hens in moveable runs complete with houses and feeders. In this way, you can boost the stocking rate to 300 birds per acre, or about 150 square feet per bird, but it is a real bore having to move them every day. I would only recommend this kind of folding as a method of fertil- ising, harrowing and reducing insect pests on land you want to improve. We did it on the grassland we wanted to dig up as a veg- etable plot. By keeping the hens on until the grass was really ruined, it made the digging much easier. We also now occasion- ally use folding to improve our orchard grass.

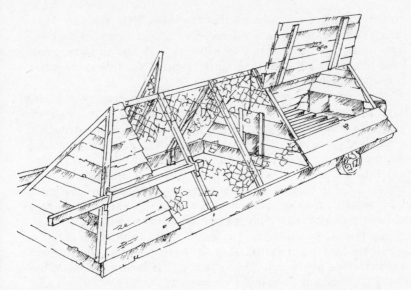

The other method of folding is to divide the grassland available into two equal runs and alternate them regularly, say once every three months. This is far less work, though it involves more investment in fencing. It also has the advantage that the hens go onto ground rich in young shoots. The method was very popular before the War and was known as 'the semi-intensive system'.

THE STRAWYARD

The strawyard is just that - a yard covered in straw. The system was developed as a way of utilising disused farm yards and cheap straw. As we saw in the last Chapter, hens are not really grassland birds. They want the dry leaves and rotting vegetation of the forest floor to scrap about in. The strawyard is a better approximation to their natural environment, in my view, than grassland range. So long as they have also got access to young green shoots, this is an excellent system. You have to add straw regularly to keep a reasonably dry top layer, but how much depends to some degree on which is scarcer for you - straw or space. If you are really short of space, you can stock them at up to two and a half square feet per bird, excluding the house, but four square feet or more is to be preferred.

At the time of writing, straw is fetching very high prices. The price of imported feedstuffs for cattle has rocketed and straw is being used as an alternative feed. This situation, is, in my view, unlikely to continue indefinitely. Straw should return to its real value-price as a litter, and will again be produced in excess quantities as more farmers turn from milk to cereal production.

The strawyard system will work using any readily available dry litter material: dry leaves from the park keeper, wood shavings, peat moss, spoiled hay, grain husks, pea straw etc. It has two magnificent by-products. The litter rots down to a splendid compost, ideal for garden use. Not only that, but rotting vegetation attracts an enormous earthworm population below it and thus the ground is generally improved. Because I have grassland unsuited to other purposes, I personally use the free range system, but I consider the strawyard system to be by far the best for people with limited space. I would even go so far as to speculate that when artificial fertilisers get even more expensive, and batteries are eventually outlawed, the strawyard could become the commercial poultry system of the future. How about that for sticking my neck out?

THE DEEP LITTER HOUSE

Deep litter is the technical term for adding on more straw or dry litter instead of cleaning it out like they used to. Deep litter houses are like strawyards under cover, but the litter is dryer and does not rot down in quite the same way. If, because of neighbours, you just have to keep hens inside houses, this is the only way that I would recommend. They need feeding more carefully and expensively of course; the housing is expensive, and you need forced ventilation really - but they do better in Winter and it can be done virtually anywhere - even in the heart of the city. You should allow three to four square feet of space per bird.

THE DIRT RUN

This is the system traditionally used by nearly all backyarders. You have all seen it in operation. The run starts off as grassland, but is overstocked and soon deteriorates to bare dirt, dotted perhaps with a few nettles that the hens will not eat, and some very dirty sink bowls, bits of car tyre or the like serving as feed and water containers. The dirt run has done more to give poultry keeping a bad name than anything else I know. It looks revolting, it smells, and attracts flies. The hens are more prone to disease and the ground is left soured and useless for years. If you are using the dirt run system at present, do not be too upset by my remarks. But read on and consider the advantages of changing to a better system.

HOW TO CHOOSE

Decide first how many hens you need to provide for your needs and then use the space you have available as extensively as you can. This means you keep your hens as 'naturally' as you can; they eat a proportion of wild foods, cost you less, are less prone to disease and give you tastier eggs.

Statistics of average production suggest that commercial free range hens lay fewer eggs than either deep-litter or battery birds kept inside. This is partly because of Winter lighting and partly the effect of Winter weather conditions. Nevertheless even if she is out all Winter, it is reasonable to expect at least 200 eggs in each of her first two laying years from a bird of good parentage - and you should get more. We in Britain consume an average of 251 eggs per person per year, but if you keep hens, I reckon you should allow for increased consumption, more like 350 per year. On that basis, here is how many laying hens you should plan for:

Number in Household	Laying Hens
2	4
3	6
4	7
5	9
6	11

Now comes the question of whether to keep a cock. Assuming that you do not want to do your own breeding, and I would not recommend it for the backyarder starting off, you do not actually need one. But they do look so fine, and it stands to reason that the hens must be happier with a male around. My view is that if you are keeping six or more hens, you will hardly notice the extra food cost of a cock, so if you like the look of them, keep one. If you are very keen on poultry meat, it can even pay to keep a cock. You buy one from a meat-producing strain and use him to breed young purely for meat, while you re-stock your layers from a good outside source.

So let us assume that you have decided how many birds in total to keep. Now, how to fit them into the space you have available? If you are lucky enough to have 100 square feet or more per bird, you can go for folded free range, using the alternating run method. If you have actually got 400 square feet per bird, you will not even need to alternate, although I believe that it is better for the land to do so anyhow. Below 100 square feet per bird, I would go for the strawyard system, except when you might disturb neighbours where the deep litter house will be preferable.

Of course, you may decide later to expand your flock so as to have eggs for sale. Fresh eggs of good quality command premium prices and can be sold from the house direct to callers once you get known. If you think you may expand later, allow for this when you choose your system.

HOW THE RECOMMENDED SYSTEMS WORK IN PRACTICE

1. FREE RANGE - THE AUTHOR AND HIS WIFE

My wife, Gloria, and I have kept chickens on free range now for about four years. We have been lucky, we have had plenty of space. When we started, I built a very simple wooden range shelter out of second-hand timber, but I am in the process of replacing it with an improved design, see Chapter Six. It can be lifted quite easily by the two of us.

Chicken wire is now outrageously expensive to buy new, although you can be lucky picking it up for less at sales. Frankly, it is now totally uneconomic to attempt to fox-proof free range chickens by fencing, (see Chapter Six on other methods), so the real purpose of my fencing is to keep the hens off our garden. That is why I have used plastic netting - the kind used for keeping pigeons off vegetables - supported on bamboo poles.

Originally, we 'folded' the run by moving the netting fence, but that is rather unnecessarily time-consuming. It is far better to fence two or more runs permanently and alternate. From time to time we confine them to a moveable 8 feet by 5 feet run to reduce grass cover before digging or to fertilise the orchard. We also keep them in there when we go away for weekends - it is reasonably fox-proof and makes things easier for our neighbour who comes to feed them and collect the eggs.

We feed them before breakfast and before dusk. They have a purpose-built poultry drinker which needs filling every three to four days. We change their nest straw every week, and clean out the hen house every month or so. That is about it.

2. STRAWYARD - LADY EVE BALFOUR

Lady Eve Balfour, a founder of the Soil Association, kept hens for a number of years on a most interesting variant of the straw-yard system. The hen strawyard was part of a productive garden rotation system. Each year, the strawyard was moved on and the earthworm-rich soil used for vegetables, assisted by the composted straw-manure mixture taken off.

Lady Eve evolved a splendid way of getting over the two main problems that hens cause on grass. The first is that they eat the young shoots and leave the older grass. Secondly, they scratch holes in amongst the grass. All in all, the effect of hens left alone in grass, however extensive the free range, is to create a very patchy and unhealthy looking pasture.

Lady Eve's approach was to allow her hens access to a small newly-sown grass plot as well as to the strawyard. Their food was always scattered in the straw, never in the grass. Consequently, the hens only went into the grass area for occasional nibbles at tasty young shoots. The rest of the time they spent scratching and pecking around in the straw. In this way, they got their greatly appreciated green shoots and the grass stayed in good condition, in spite of being stocked at a rate of around 3000 to the acre, or 15 square feet per hen. I consider this a real breakthrough in natural chicken-keeping techniques, and it amazes me that no one else has yet taken it up.

Lady Eve assisted the survival of the grass plot by dividing it and alternating it in the usual way. She also kept the plot well mown to encourage plenty of young shoots. The strawyard was added to with fresh straw as needed and cleaned out completely every twelve months.

For those of you with a pioneering spirit, I am going to suggest a few further variants on Lady Eve's system. Firstly, how about rotating a hen strawyard with mulch vegetable gardening? Why go to the trouble of digging up the year old rotted straw and manure mix? It might well make an ideal mulch left in place, while the chickens are moved on to a new plot. For more on the amazing low work mulch gardening system read 'How to Have a Green Thumb without an Aching Back' by Ruth Stout, published in the U.S.A. by Cornerstone Library, which either your library or your bookshop should be able to get for you.

Secondly, how about building the grass plot into the rotational system, so that the hens are always eating freshly sprouted shoots? Freshly sprouted grains and seeds have incredible properties - that is why they are eaten by health food enthusiasts. For instance, the vitamin C content of wheat increases by 600 per cent in the early stages of sprouting. The vitamin B2 content of oats increases by 1350 per cent. Old rats fed in an experiment on sprouted seeds actually became more youthful in appearance and activity - and so the evidence goes on. The odds are that hens having access to newly sprouted seeds and grains will also gain health advantages. As soon as I get round to it, I intend to try out these ideas - maybe you will too.

In principle, I figure the rotation, starting on undug grassland, would be:

First year: Hen strawyard, kept on until rotted depth at least six inches.

Second year: Vegetables under six inch mulch.

Third year: Vegetables under six inch mulch.

Fourth year: Vegetables under six inch mulch.

Fifth year: The unrotted top layer of straw is removed and a herbal ley seed mix is sown on the rotted mulch, a strip at a time through the year. The hens are kept off the growing seeds until they are fully sprouted; then they are allowed on.

Sixth year: As first year.

It will need some experimenting to find the hen stocking rate which will result in a six inch rotted depth in twelve months. If you are very limited in space, you could alter this from two years hens and three years vegetables to two years hens and four, five or six years vegetables. I reckon that there would be so much goodness in the soil and the rotted mulch after one or two years of hens that it would easily grow six years of vegetables with nothing added except possibly some lime. This is all suggested on totally untried speculation, but if it works, it could turn out to be the ideal way of utilising a small garden plot for maximum combined production of eggs, vegetables and a little poultry meat with no digging, no bought fertilisers, no compost heap, no spoiled grassland and no smelly dirt run. That is why I feel justified in including it.

3. DEEP LITTER - RICHARD AND RUTH PAINE

The Paines live in a semi-detached house very near the centre of Dorking - a very urban setting for poultry-keeping. The drawing shows the interior of the garden shed which they use for their miniature deep litter system. Half the shed is used for garden tools, bulk foods and a freezer, leaving an area 7 feet by 10 feet for their sixteen hens. The area is fenced off with chicken netting

plus a ten inch lip at the floor to contain the built up litter. Inside are a nest box and a broody box, both wall mounted to give more floor space, plus a trough feeder and a drinker. The shed door and windows are left open most of the Summer. There is no forced ventilation.

The birds are Golden Comet hybrids. They are fed on proprietary ready-mixed Layers Mash plus a Grit and Oyster Shell mixture. Until the price got ridiculous, the Paines used straw for litter. Now they use peat moss, but find it very dusty, particularly for a hut used for food storage as well as chickens. The litter is cleaned right out once a year. All garden wastes, kitchen scraps etc. are fed to the hens as well. Eggs are sold regularly to friends and neighbours at marginally less than the local farm market stall price. In two years they have had no complaints from neighbours as to noise or smell. The Paines find urban hen keeping a very sociable hobby. Children come to see the hens; word gets round on the quality and freshness of the eggs, and friends have been encouraged to start for themselves.

In Winter, the hens are persuaded to lay more by an artificially long day. By the use of a time switch on the shed light, the day-length never reduces below 17 hours, so the birds are never triggered off lay by the pattern of reducing natural day length. While we are on the subject, it is better to give the extra light in the early morning. If the light switches off suddenly in the evening, the birds may be unable to roost. For details of other more complex systems of artificial lighting see "Lighting for Egg Production", Ministry of Agriculture Advisory Leaflet No. 540.

At present price levels for compound foods and straw, Richard Paine doubts whether he has saved much using this system. If he had more garden, he would prefer to be able to let them outside. Nevertheless, the Paines thoroughly enjoy their miniature urban poultry business.

4 which hens to buy

Hens can be bought from various sources and at various stages of maturity:
* from a hatchery as 'day-old chicks', and they are usually literally that age;
* from a rearer at 6, 12, 18 weeks or 'point of lay' which should mean around 20 weeks, but can mean somewhat less;
* from a pet shop or poultry market at any age;
* from a battery producer at about 76 weeks, sold off having finished their first egg-laying year.

The first important point to remember about choosing your source is this: the essential reason why hens lay on average seven times what a wild fowl could achieve is that they have been carefully and continually bred for egg production. The biggest single factor affecting how many eggs you get from your hens will be their inbuilt genetic potential as layers. Bearing in mind that the cost of a chick is tiny compared with the food you will eventually buy to feed it, there is absolutely no justification for buying inferior stock. My advice is to avoid pet shops and markets altogether - unless you later decide to breed and become capable of judging poultry by examination.

It is also important to realise that hens are not particularly adaptable creatures. They always do best if they are reared to maturity under the same conditions as you use for them in later life. A bird reared inside in deep litter or batteries, as so many are commercially of course, will not generally adapt well and quickly to life in an outside run. Hence my second piece of firm advice - do not buy reared stock unless you are quite sure that they have been reared to suit your conditions. In practice, that is often difficult to be sure of, so I am not greatly in favour of buying reared stock at all, in spite of the fact that it means you start getting eggs sooner.

As to ex-battery birds, they are a very cheap and reasonably quick way to get into egg production, (they should come back into lay 8-10 weeks after purchase). But they too tend to have difficulties adapting. It is painful to see them slowly learn to walk, peck at grubs, scratch and perch after a life of imprisonment from birth. They also have to grow decent feather-cover. After all that, to expect much out of them in egg production is asking a lot. You are almost bound to lose some in the early stages; you may bring in disease problems and when you do come to eat them, they may still have the sores and lumps associated with battery life.

My advice is to buy day-old chicks. They are the cheapest way in: you have the enjoyment of rearing chicks right from the start; they are reared to exactly your conditions and they get to know you. The only drawbacks are that you have to wait five months for eggs, and that you need either a broody hen or a home made artificial brooder to keep them warm, (see Chapter Seven).

WHICH BREED ?

If you want the maximum number of eggs for the minimum bought feed cost, then you should go for Khaki Campbell ducks, which lay 300 eggs per year or more and live mainly on grass. If so, turn straight to Chapter Eleven. If you prefer to stick to hen eggs, then the first question to decide is whether to go for a white egg layer, or a brown layer. White egg layers lay slightly more eggs per year and eat slightly less food. They are generally also less prone to broodiness. But in favour of brown eggs, they fetch better prices when you sell them, (not only that, but some customers do not actually believe that white eggs can be free range). Brown egg breeds are also less flighty and highly strung than the smaller white egg layers. They also look more like the traditional hen. And let's face it, a bit of broodiness now and then is no bad thing in a backyard hen. I only know one backyarder who keeps white layers. Most of you will go for brown eggs, I'm pretty sure.

Whichever colour egg you choose, there is still a choice to be made from three totally different breeding systems. Should you buy a pure breed, a first cross, or a modern hybrid? Virtually all commercially kept birds are hybrids, but a lot of backyarders

THE WHITE LEGHORN

and commercial free rangers still keep pure breeds or first
crosses. Please bear with me while I attempt to explain what
this is all about.

The most common pure breed of white egger is the White Leghorn. This was bred in Italy and imported into the U.S.A. back
in 1835. After a long process of selective breeding, it is now
known as 'the egg machine'. But this breeding for egg quantity
left out lots of other important factors. It is a very highly strung
bird, difficult to keep in. Some strains have other weaknesses as
well.

The most important pure breeds of brown egger are the
Rhode Island Red and the Maran. The Rhode was bred in the
U.S.A. around 1880. It is bigger and more docile than the Leghorn; a fine red or gold feathered traditional-looking hen. In
spite of averaging slightly below the Leghorn, a Rhode still holds
the World laying record (361 eggs in 365 days). The Maran is
French bred, and not a great quantity layer, but it lays the most
beautiful dark brown suntan-coloured eggs. It is also supposed
to lay rather better in Winter than other pure breeds. It has a
fine mottled grey feathering and is popular with backyarders, but
tends to be nervous and shy until fully settled in.

Breeding is a very complex subject, much tied up with Mend-
elian genetic theory and the mathematical laws of probability. I
shall try not to bore you with more than you need to understand
and evaluate what type of hen to buy. Creating new strains of
poultry and improving them inevitably involves inbreeding and
can lead to weaknesses developing. It was discovered very early
on that if you crossed two inbred pure hens with good laying records, the resultant offspring could, in certain cases, retain
these laying characteristics, but also have the advantage of more
vigour (known as 'hybrid vigour'). But this only works once. If
you breed from these mongrel offspring, the next generation will
be very mixed, in looks and in performance. Early breeders
were not slow to spot the commercial significance of this fact.
People who buy first-cross chicks have to keep coming back for
more, unless they are prepared to keep two lines of purebreds
just for breeding.

Certain first crosses also exhibited another feature of great
commercial significance, discovered in 1921. Even at birth, the
male chicks were a different colour to the females. Hence day-
old chicks could be sexed six weeks earlier than previously. The
main surviving example of the first cross is the Light Sussex/
Rhode Island Red cross. The Sussex was originally bred as a

THE RHODE ISLAND RED

table bird, though its laying properties have since been improved. The cross is sex-linked, the male being white and the females orange-brown. It produces a good layer, and a pretty decent table bird, hence a very popular all-round bird for backyarders.

The dramatic breakthrough in chicken breeding has been hybridisation, a technique of breeding developed in the U.S.A. between the Wars. If you breed by the traditional method of selective breeding, you attempt to improve a pure strain by eliminating the poorer performers each generation. Suppose you start off with a spread of laying records over a year of between 150 and 200. By selection you can improve the average gradually from 175 to 180 and maybe 190, but there is an inevitable plateau effect with this system. You cross strains and thus establish new, better strains, but that is about all.

The principle behind hybridisation is that if you breed a first cross between two pure strains, the resulting offspring does not necessarily perform somewhere between the two. You may get what is called a 'nick', where the offspring do worse than both, or better than both. If they do worse, you chuck them out, but if they do better, you have created a breeding breakthrough, a modern hybrid.

Just to show how this has affected laying performance, here are some average laying records from the U.S. Dept. of Agriculture trial flocks kept at Beltsville, U.S.A. The figures are for 52 weeks in lay.

Improvement in Average Laying Performance

<u>1932-36</u> (the days of pure breeding and ordinary non-nick first crosses)

Rhode Islands (open line continually improved from the best of American outside stock)	192
Light Sussex/Rhode first crosses	172

<u>1937-45</u> (when they continued to improve the pure lines and also tried keeping some closed)

Rhodes (open line)	189
Rhodes (closed line)	173
White Leghorns (closed line)	195

<u>1946-54</u> (when they tried 'nick' hybrids)

Rhodes (open line)	217
White Leghorns (closed line)	212
Inbred Rhode x Inbred Leghorn (hybrid)	233
Inbred Leghorn x Inbred Rhode (hybrid)	258

In 20 years of gradual improvement by traditional means, they got an improvement of 25 eggs per year, of which they assessed half was due to improved feeding techniques. By hybridisation, they got an immediate improvement of 30 eggs per year over the purebreds.

You will find that quite a few backyarders are dead against hybrids. I myself was very undecided until recently. If you keep purebreds, you can do your own breeding and even gradually improve the performance of your flock by selection. If you keep hybrids, you have to go back to the hatchery for replacement stock. Not only that, but hybridisation has concentrated the breeding industry into very few hands. The reasons are that breeding has to be done on a big scale now to be competitive and that breeding hybrids may sound simple, but creating really good hybrids is not easy and it is only the best firms that have survived.

About a dozen firms dominate the poultry breeding business worldwide and nearly all of them are American. At least three of them are owned by pharmaceutical companies, which does make one wonder whether the hybrid of the future might need big doses of antibiotics as regular protection against disease. I have to admit that I do not like relying for new stock on these people. However, I accept that hybridisation, carried on in a really big way by international organisations, has greatly improved the genetic egg-producing potential of the modern domestic fowl. So long as they continue to produce hybrids which look reasonably attractive and adapt to natural systems of poultry-keeping without any increased sensitivity to disease, I shall continue to buy from them and I recommend that you do too.

The only possible disadvantage you have to accept with hybrids is a reduced tolerance of poor management. Hybrids are highly bred to perform outstandingly. But if you fail to keep up their protein intake, or forget to keep the drinker clean, the results are more likely to come faster and more seriously than with the less 'highly tuned' traditional breeds. However, I personally regard this as a challenge rather than a disadvantage.

It is very difficult to make definite recommendations as to which particular hybrids are best for natural systems. All the various laying tests carried out in different parts of the world are now based on battery conditions, so that is the only guide we have. In the U.K. tests (carried out by National Poultry Tests Ltd.), the big success in the brown egg league is the Warren-Studler SSL, which came first in four out of the last five tests (a flock average of 286 in 56 weeks in the last test). The Kimbrown has a good test record, but is said to be losing ground to newer hybrids, such as the Shaver S585. Other good brown-eggers are the Arbor Acres SL, the Hi-sex Brown, the Babcock B380 and the Hubbard Golden Comet.

Most of the firms named also have good white-eggers, but the leaders are reckoned to be the Shaver S444T, the Hi-sex White and the Babcock B305.

The addresses of the U.K. headquarters of these firms are given below. Most of them operate through networks of independent local hatcheries and will supply you, on request, with the name and address of your nearest one. Alternatively, look up 'Agriculture, Ministry of' in your local phone book; ring them and ask for the ADAS regional poultry adviser - he should know all about local hatcheries.

> Warren-Studler Breeding Farm Ltd., Orton Longueville, Peterborough, Hunts.
> 'Kimbrown' is by Kimber Breeding Ltd., Braintree, Essex.
> Shaver Poultry Breeding Farm (GB) Ltd., Bawdeswell, Dereham, Norfolk.
> Arbor Acres (UK) Ltd., Rough Hill Farm, East Hanningfield, Chelmsford, Essex.
> 'Hi-sex' is by Eurobrid Ltd., 50-58 Pensby Road, Heswell. Merseyside.
> Babcock Farms Ltd., Cambridge.
> Hubbard Ireland Ltd., 54-56 London Road, Stroud, Glos.

NON-HYBRIDS

As you see from the above, I am convinced that the advantages, in terms of eggs per £ of feed cost, of using hybrids outweigh their disadvantages. However, I can see that this is still very much an issue of debate amongst backyarders. Self-sufficiency enthusiasts in particular find it difficult to match reliance on multinational breeding corporations with their principles.

If you prefer to steer clear of hybrids, you can either go for first crosses, the most popular of which is the Rhode/ Sussex, or for a pure breed like the Rhode, the Leghorn or the Maran. There are still a few small regional breeders around who can supply either chicks or growers of this type. In most areas, it should be fairly easy to track down a source of Rhode/Sussex crosses, but pure breeds of good laying quality may be more difficult to come by. Try the Yellow Pages, or ask other 'free range' poultry keepers in your area for good sources.

Even if buying non-hybrids, my initial remarks at the start of this chapter still apply. Go for day-olds if you can, and take some trouble to get chicks with good genetic laying potential, that is to say that they have been bred carefully by capable poultrymen. Be warned that breeders vary considerably in quality and reliablility. Get advice and go and see the place to form your own impression. Be cautious of claims for 'new breeds' from small independent breeders - they are unlikely to have the capabilities to produce reliable new hybrid strains.

5 what to feed them

As we saw in Chapter Two, the natural feeding sources of the jungle fowl are the seeds, young green shoots, grubs and worms of the forest floor. My approach to the question of what to feed domestic fowl in confinement is to attempt, so far as possible, to cater for their natural preferences and to feed the foods that their digestive systems are best designed for. Nevertheless, the only one of these natural food sources which can usually be supplied in adequate quantity on the ground already is young green shoots. Except on extensive range in summer, the supply of grubs and worms will need to be supplemented in some way from elsewhere. Seeds will have to be supplemented throughout the year whatever the system.

It is pretty impractical to grow your own supplementary supply of grubs and worms for feeding the hens, though perhaps it could be done by folding them on areas of rotting vegetation or feeding them trays of worms reared on compost. The best cheap alternative sources of the protein, calcium and other nutrients that hens get from grubs and worms are 'blood and bone meal' (crushed up slaughter house wastes, sold as fertiliser by your local seed merchant - buy it loose, not in branded packs), or 'fish meal', also sold as fertiliser. The best cheap alternatives to wild seeds are whole cereal grains like wheat, barley, maize, oats etc., though other smaller seeds have been used in the past when they were reasonably cheap (sunflower seeds, linseeds, hemp etc.). Any of these grains can be used, but the domestic hen's order of preference is reckoned to be:

wheat, maize, rye, barley, oats. Some people believe that barley is no good for laying and makes hens go fat. It is true that it has the lowest protein content of these grains and that may be the reason. The answer is to allow the hens a little more blood and bone meal to balance things out.

Now of course, the domestic hen is capable of utilising just about any waste food that you have around: kitchen scraps, garden wastes, skimmed milk or food factory wastes of all types. But it is important to realise that the hen's system is not designed for coping with high volumes of 'bulk' foods with high fibre and low nutrient content (unlike the cow and the duck). If you fill her up with this type of food, the chances are that she will lose out on one or more of the important nutrients and her health will suffer. For those that feed a high proportion of kitchen wastes, commercial 'balancer rations' are available, but frankly, if it is an animal for converting wastes to eggs you want, get Khaki Campbell ducks and get used to duck eggs.

If you do feed kitchen scraps (and most people do, but in limited quantities), avoid citrus peel, salt and small fish bones. According to official regulations, The Diseases of Animals (Waste Food) Order, 1973, any meat-based waste should be boiled at 100 degrees C for 60 minutes to stop the spread of animal diseases. If you have already cooked the meat for your own purposes, I cannot see much wrong with feeding the wastes unboiled, but that is your problem. Those are the rules.

To sum up, the best natural diet for a domestic hen consists of:
* as much access as possible to young green shoots;
* as much access as possible to grubs and worms, supplemented as necessary by blood and bone meal, or fish meal;
* whole grains or seeds;
* fresh, clean water;
* sunshine.

This recipe is practicable if you use a free range or straw-yard/grass plot system, but not if you keep hens inside on deep litter. In that case, you either have to bring greens in to them or go away from the natural over to the modern analytical approach to feeding. That means feeding ready-compounded 'layers mash', presumed to contain the right balance of measurable nutrients like protein, vitamins etc., but of course in a totally unnatural form. I do not like this approach, but I have to admit that, as yet, I know of no-one who has tried the natural feeding approach in deep litter conditions, so I feel obliged to add the analytical approach as a method of feeding. Perhaps by the time I do a second edition, some of you will have written in telling of results with natural feeding on deep litter.

If you do decide to try it, remember that the house must allow the birds sunlight. Use whatever greenfood you have available nearest to young green shoots. As to quantities, it is difficult to be precise, but laying hens will consume up to 2 oz of greenstuff per day if it is readily available.

Some people, notably the Henry Doubleday Research Association, are very keen on growing Comfrey for feeding to livestock. It is certainly a very nutritious greenstuff with a big yield per square foot and as such pretty useful in deep litter conditions where space is short. However it is only good in Summer, when garden wastes should normally be adequate anyway to cover most of the hens' greenstuff needs. It is in Winter that you will need specially grown feed and for that purpose, Kale takes a lot of beating. What I do not agree with is the idea of feeding Comfrey or any other greenstuff in such quantities as to attempt to replace either the seed or animal elements of the hen's diet. (Kale can be grown from seed, but Comfrey is usually started from cuttings available from the HDRA, Convent Lane, Bocking, Braintree, Essex.)

HDRA member Mr F. Roscoe kindly supplied me with information on his experience of Comfrey feeding. He fed Comfrey as about one third of his hens' total diet and reports no problems, although the flesh of the birds when killed was slightly yellower than usual. He recommends planting up to 5 Comfrey plants per bird for a really good supply through the Summer.

FOOD QUANTITIES

A domestic fowl producing a 2 oz egg or bigger virtually every day on a total food intake of around 4 oz per day clearly needs to be carefully fed. They are fairly stupid birds and will peck at almost anything edible. The capacity of the hen's crop is, however, only 4 oz, so if she fills herself up with relatively useless foods she will soon have no room for more and will end up undernourished.

On a natural feeding system, balancing the diet can be tricky. There are various approaches. Perhaps the simplest is the total ad-lib system developed by Jim Worthington, author of 'Natural Poultry-Keeping'. You supply unlimited quantities of grain and animal meal in separate containers and let the birds balance their own diets between these two and the wild foods on the ground (shoots, grubs etc.). My feeling is that that was alright in the days of cheap imported grains, but not now. I am convinced that it makes the hen inevitably less bothered to look around for wild foods and so works out more expensive. It also obliges you to construct special sparrow-proof feed hoppers.

What are the alternatives? How else do you balance the hen's diet? Some people do it by time, giving them access to grain and animal meal hoppers for 20 minutes twice a day. Some do it by measuring out standard 'approved' quantities. For instance, they feed 2 oz of grain twice daily plus one third of an ounce per day of animal meal.

Whichever way you do it, how do you judge whether you are getting it right? More especially, how do you judge that you are not discouraging free wild food consumption by a slight over-supply of easy-to-get bought foods? My approach is based on the idea that you can judge best on a small scale by observation of three factors:
* how your egg production compares with what it should be
* the weight of the hen
* the fullness of the hen's crop an hour or more after she has roosted for the night

Typical production rates for hens kept outside are given in Chapter Two. If your hens fall badly below these rates, check hen weights and crop fullness to see if the reason is under-feeding. The weight of a fully-grown layer varies from breed to breed, but most hybrids top the scales at 3.6 - 4 lb at point of lay, rising to 5 lb at the end of the first laying season. Find out from the hatchery what your hens should weigh and check the average. One way to weigh a hen is to put her inside a sack. That keeps her still long enough to get the weight.

Now the crop should, it is said, be at least half full (i.e. 2 oz) on roosting. It should also be half full during the day when the hen rests for a preen. If you fill a plastic bag with 2 oz of grain, it gives you a standard against which to judge the crops you feel. The hen's crop, by the way, can be felt very easily just below her neck. It is a loose bag and its fullness is pretty easy to judge.

These three factors give you a method of assessing whether you are underfeeding. But should you measure what you feed by time or weight of food? Doing it by time allows the birds to balance out their own grain/meal ratio, but on the other hand, doing it by weight enables you to cast the food around to encourage scratching. On balance, I prefer to do it by weight, for the reason that a slight overfeeding of animal meal will not work out too costly, but lack of encouragement to scratch will.

Remember that the supply of wild foods is very seasonal. In the Winter, I assume that at least 3 oz of the total has to be in the form of supplied grain and that a full one third of an ounce of animal meal needs to be fed (a little more if you feed barley or oats). Your birds may need much more – it depends upon your land and their keenness to scratch. In Spring, we reduce the animal meal gradually to nothing, because they have plenty of grub-catching range. The grain is also reduced slightly, but with a watch on production and crop fullness. We find that production begins to suffer under about $2\frac{1}{2}$ oz.

The daily requirements of animal meal are so small that we prefer to feed it twice a week. Six hens get a twice weekly ration up to 7 oz fed to them in a stable dog bowl straight after their morning grain feed. If you feed it moistened, it is more palatable.

WATER, GRIT AND CALCIUM

Apart from their main nutrient requirements, hens also need access throughout the day to clean, fresh water and lots of it. Eight hens get through half a pint every day. They also need a source of insoluble grit with which they refill the grinding stones of the gizzard. That is no problem on range, but if you suspect a shortage, a small box of grit in the run will be needed – to peck at when they feel the need.

One more thing needs watching. Hens producing eggs regularly need lots of calcium for shell building. If you are on chalky soil, there should be no problem. They pick up enough chalk with their food. If you notice, however, that the egg shells are on the thin side and tend to crack easily, you will need to supplement a calcium source. The cheapest source is used eggshells but do bake them and crush them up first, just in case the hens get a taste for fresh eggshell and the clinging eggwhite, in which case they may start eating your egg supply. The other source is crushed oyster shell, which is commercially available from poultryman's suppliers and grain merchants.

RECEPTACLES AND STORAGE

Food and drink receptacles are important. Badly designed ones cause great waste of food and encourage rats. If they are not kept clean and moved around the run from time to time, they can be the cause of spreading disease quicker than anything. For

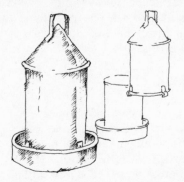

drinking water, I recommend the Eltex design - but buy one which will last for 4-5 days, no bigger. It is a great design, easy to fill and carry and, very important, the hens cannot excrete into the water.

You will need a receptacle for feeding protein concentrate. They eat so little that feeding it twice a week is adequate. We use a dog-bowl which they cannot tip over and is big enough for all the hens to get at. It is removed and is washed after every session. Grain should be scattered, in my opinion, although you can buy sophisticated grain hoppers if you so wish.

The grit and shell container can be a dirt collector and it helps to mount it high so that they do not walk on it. It should be of such a design that they cannot perch on it or exrete into the grit. Make your own or use the Eltex grit hopper. It is not that I have shares in Eltex but rather that they seem to have a virtual monopoly in the supply of this kind of equipment. Luckily, their stuff is good.

For deep litter feeding of mash, special Eltex hoppers are available. Greenstuff is best given off the floor, either hanging down in bunches or held by netting or wire spaced so that the hens can reach through and peck at it, but so arranged that greenstuff pulled down is not wasted.

If, like us, you tend to go off for weekends but lack neighbours who are willing to feed your hens and collect the eggs, then the Eltex feed hopper, left open with two days' rations for ad-lib feeding is the answer. If they are not used to it, they will over-eat on the first day, but do no great harm to themselves.

Fairly obviously, you will need to make some provision for storing the grain and other bulk foods your hens will need. The more you buy at a time the lower the price becomes, but it is only worthwhile if your storage is good. Grain must be kept dry and away from rats or mice. We have had problems of grain deterioration, mainly attack by little borer insects, so now we forget the discount and store only three months supply. Plastic dustbins make fine storage containers, although old half-cwt industrial containers made of plastic, metal or board are cheaper if you can track them down. If you suffer from rats, metal containers are the only certain method of storing grain safely.

ADDENDUM FOR STRICT VEGETARIANS

The domestic fowl is not by nature a vegetarian. However, those readers who are themselves strict vegetarians may wish to keep hens without having to resort to blood and bone meal or fishmeal, both by-products of human flesh-eating habits. If you prefer, you can feed protein in vegetable form, though it will probably work out more expensive. The most concentrated vegetable source of protein is the soya bean. Hens will eat it in various forms: as whole beans, soaked to soften them or grits, soaked or unsoaked or as flour (grits are rough milled beans). These forms have to be bought from whole food shops at high prices. Alternatively, ask your grain merchant for soya bean cake or meal, sold as fodder.

The HDRA suggest that daffa beans be used instead of soya beans as a protein source. Their main advantage is that they can be grown in Britain - they are also hardy and yield well. They are too large to be taken whole by the birds and need to be coarsely milled in a hand grinder.

6 confining and housing them

So far, what I have said about systems of management, breeds and feeding is unlikely to be disagreed with by other poultrymen - I have tried to give a variety of options. In this chapter, however, some of the options I propose are pretty personal ones and some other poultrymen will think them absurd. I will try to continue putting forward various options - take your pick.

WHY CONFINE?

Poultry have traditionally been confined by fencing for two separate reasons. Firstly, to keep them in and secondly, to keep predators out. It makes very good sense to keep chickens off vegetable plots, flower beds or field crops, all of which they will take great pleasure in ruining. So even if you want to give your hens relative 'free' range, it is advisable to separate them from such areas, either by fencing your garden or fencing your hens.

As for fencing to keep out predators, the cost of truly pre-dator-proof fencing is now so astronomic that I am personally dubious of the economics of attempting much more than deterring predators, as opposed to proofing against them. But before I go into that one, it is as well that you know what problems I am talking about.

Public Enemy No. 1 is the fox. In my opinion and in the opinion of many countrymen, this animal should be classed officially

43

as vermin and the preservation of foxes for hunts made an offence. Unfortunately the population of foxes seems to be increasing rapidly with the disappearance of their main enemy, the gamekeeper. Foxes are now regularly sighted even in the centre of large towns.

The fox is an opportunist. He hunts and eats whatever works out easiest and least risky for him. The hungrier he is, though, the more trouble and risk he will take. Foxes are very partial to all poultry - at night chickens offer virtually no resistance and can be killed in great quantities on their perches, not just for food but for the hell of it. In the early morning, before their owner is up, a hen is pretty easy to catch on the run.

Public Enemy No. 2 is the badger, but there are few about and they are generally not keen on approaching houses. The badger hunts at night only and usually kills and takes only one hen.

The other important predator of fully grown hens is the stray dog in towns. Country dogs, brought up near hens, rarely go for them. Strays can be a real nuisance and your only consolation is that you can claim full damages against the dog's owner (Dogs (Amendment) Act, 1928), that is if you can trace him.

Chicks have many predators, including crows, rooks, magpies, stoats, weasels, rats and even cats. The mother hen helps to protect them if she can, but the real answer is to keep them in a small predator-proof run until they are about six weeks old and over the main danger.

WHAT FENCING?

To achieve virtually 100 per cent predator-proof fencing, you

44

need to go to chain link fencing, as a really keen fox or badger will bite through ordinary chicken wire. It needs to be at least 5 feet high so a fox or dog cannot jump it, with an overhang outwards at the top so he cannot climb it. It also needs to be let into the ground to a depth of at least 1 foot so he cannot dig underneath it. The cost of the chain link alone on such a fence works out at £1.20 per yard length. Then there are the 6 ft posts, overhang arrangement, tensioning wires and digging. You can now see why I consider it uneconomic, even for a limited strawyard/grass plot area.

Accepting that fencing can only deter and not stop predators, I recommend concentrating attention on making the night-quarters predator proof and keeping hens inside or in a very limited safe run in the early morning before you get up to let them out to range. I go for the cheapest chicken-proof fencing available, namely black plastic fruit netting. Six foot high, it works out at 14p per yard length. I say black, meaning the slightly stronger square mesh type rather than the green diamond shaped mesh variety, which tears too easily. For adult hens only, you could alternatively use the brown square mesh pigeon proofing netting. It has a larger mesh and is even cheaper. Plastic netting can be held by bamboo canes or other bendy sticks. Ones that bend rather than snap make the fence more of a deterrent to predators who attempt to climb it. To deter digging, either dig it in to a depth of 1 foot if you can be bothered, or leave 1 foot overlap at ground level and weight it down on the outer side with stones, bricks or small logs. Any would-be burrower gets in a tangle.

The only thing that you are not safe against is biting through by the predator. If you live in a fox or badger area, you can back up the netting with 3 feet high sheep netting placed immediately behind, bringing the fencing cost up to 30p per yard length, still a quarter of the price of the full chain link arrangement. Alternatively, the netting can be backed up with a couple of strands of electric wire at 6" and 12", which should stop biting through quite effectively. I should mention that we have not had predator problems where we live, so I only have experience of the netting alone. Having read many suggestions for fox-proof fencing, I put these composites forward as the cheapest ones likely to work.

STOPPING PREDATORS AT THE HEN-HOUSE DOOR

Old books on chicken-keeping are full of suggestions for fox-proof devices to stop foxes getting to the hens at night. The surest answer is to close the door every night after they roost. But that is a bore, particularly in early Summer when they do not roost until after 10 o'ciock. Not only that, but you are bound to forget one night and that will be the night when the fox will strike. If you go away for a few days, finding someone to scatter feed and collect eggs is not too difficult, but getting them to close the hens in after dark is much more of a problem.

Chemical 'fox deterrents' are available, but it appears that some foxes, cunning as they are, have learned to associate the smell with available poultry, so the object is defeated.

Various keen poultrymen have tried to design special doors which hens can get through but foxes cannot. Not one seems to have turned out as being really effective. Others have hung 6" of iron chain centrally in the doorway so that the fox suspects a trap and stays out. My feeling is that it would depend upon how hungry the fox is. A bend in the entrance is another idea. The short hen walks round it easily but the long fox cannot.

Of all of them, my favourite is the late Newman Turner's idea of positioning the henhouse door reasonably high, say 18 inches off the ground, with an ordinary domestic metal foot scraper positioned in front. The hens can fly up and walk in, but the fox cannot stand walking on the sharp metal slats. Newman Turner claimed that it worked well over many years in a district where foxes were seen regularly.

The main danger is at night and a fox-proof entrance based on these ideas should solve that, but what about the early morning when the hens are out feeding and you are still in bed? At night the fox often kills all your hens, but in the early morning he tends to kill one and take her off, so the loss is much smaller. My answer is to have a fox proof hen house entrance for when we are away, but also to have a small close-mesh wire netting run right next to the hen house. It would take a fox quite some time to bite through the net or burrow under, so I am satisfied that the early morning danger is much reduced by shutting the hens into this small run in the evening, and letting them out when I get up. I prefer this to keeping them inside when they would much rather be out trying to catch the early worm. If you wish, such a run can be made 100 per cent predator-proof by using chain link fencing for the sides and roof, and either weldmesh or wooden slats for the floor; but then the hens would not be able to scratch, of course.

WING-CLIPPING

Hens can fly, so if you are fencing them, either to protect them from predators, or to keep them from doing damage outside, you had better make sure that they cannot fly over. The answer is wing-clipping. By clipping the first three or four flight feathers on one wing tip back to half length, the bird's flying balance is sufficiently disturbed to prevent her from getting more than a few inches off the ground. She can then be confined, if she is of a heavy breed, by as little as three feet high fencing, though higher is to be preferred. With lighter breeds, it is as well to fence to five or six feet high, because even when wing-clipped, they can fly up to a decent height.

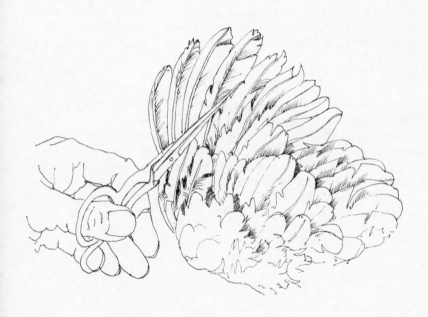

WHAT HOUSING IS NEEDED ?

BASIC REQUIREMENTS

You will probably have seen many very different-looking designs
of hen house, some of them illustrated here. This may have con-
fused you, as it did me, as to what a hen house is _really_ required
to do. So let us start from basics.

A hen house should be constructed to provide protection from
wind and rain. So, first of all, it needs a solid waterproof roof.
Roofing felt fixed over a slatted wooden roof is the usual arrangement, though exterior grade ply; vertically set tongued and grooved
boards on a well sloping roof; or even corrugated iron work well
as alternatives. The walls should also be weatherproof, with
tongued and grooved boards being the most popular solution, and
again exterior ply as an alternative.

Secondly, the house must be well ventilated. Hens excrete
half their daily quota while roosting, so if you do not allow for
ventilation, the place can get very unpleasant. The usual arrangement consists of wire netting 'windows', spaces under the roof,
setting the house on legs with a weldmesh or slatted floor, or
combinations of these.

ROOSTING NEEDS

The house will be used for two activities: roosting at night and
laying eggs during the day. For roosting, you can use three inch
by one inch weldmesh, or wooden slats. If I were a naturally
free roosting bird, I would prefer wooden slats. Two inch by
one inch timber makes fine roosts. If you are close spacing the
slats to stop predators getting through from underneath, it helps
to plane them to the shape shown. This stops the droppings from
becoming lodged between the slats.

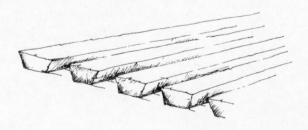

The other system is to space them wider, say 6 inches apart, and protect the hens from below with chain link, chicken wire or weld mesh. In both cases, each bird needs about 7 inches of roosting width on roosts at least 12 inches apart.

The outdoor systems recommended in Chapter Three, namely free range or strawyard/grass plot, involve moving on the hen-house to a new spot from time to time, primarily to reduce disease contamination. This also has the advantage that you can use an un-floored hen house, letting the roosting droppings fall through to the ground below. Otherwise, the house has to be designed with floors that pull out or are somehow cleanable.

LAYING NEEDS

Now to the other activity: laying. The usual way of providing a soft-floored, dark, cosy place for the hens to lay is to use nest boxes. These are simply boxes, open on one side, at least 16 inches square and at least the same in height, lined with a soft, dry litter such as straw, and so placed that they are dark and secluded. One box will serve the laying needs of up to five hens without any need for queuing. It is a good idea to place the boxes lower than the roosts. This helps to discourage young birds developing the habit of sleeping in the boxes. It is also a good idea to have a three inch lip at the base of the open side as hens are rather fond of rearranging their nest litter. The lip stops it being thrown out of the box in the process.

Although 16 inches square is the standard size given in the literature, I have noticed that when hens lay in an odd box of straw, they make a lovely neat round nest about 20 inches across, so I would reckon that something more like 20 inches square perhaps makes them happier and less likely to throw the litter out in their nest building. I shall try it next time I rebuild.

Egg eating is a nasty habit that can develop, but keeping the nest boxes really dark inside is one way of preventing it. The easiest method is to hang sacking or other material over the openings with long vertical slits to let the hens get in and out. The other preventive measures are to stop thin shells, when they occur, by feeding oyster shell and also never to feed un-baked eggshell. (See Chapter Five on Feeding).

HOUSING FOR REARING

For hatching and rearing new chicks, it is necessary to have a separate nest box, plus a small predator-proof, chick-proof run attached, at least 4 feet by 2 feet, preferably larger. The run is easy to construct from three quarter inch (or closer) chicken wire. As before, you can either make it with a floor, or make the nest box with a door to close up at night. I prefer the latter in order to give the mother hen (or 'broody') ample opportunity to teach them how to scratch for food at the earliest age.

Once the chicks are about six weeks old, the mother hen can be put back with the other layers to earn her keep again. But the 'growers', as they are then called, are better kept apart from the layers for another twelve weeks or so if they are in restricted conditions, because otherwise they get bullied and pecked off their fair share of food. On free range, it matters less as they range away from the others and make up for it with wild foods - in fact that trains them rather effectively to be good wild food users. But on restricted range you will need seperate accomodation for the growers. The essentials are roosts and a run, but both should be as similar as possible to their eventual adult home.

The arrangement I like for growers is a small run adjacent to the main hen house, which itself has a small 'growers only' door out to it. The growers sleep with the layers, but feed seperately. The opening should be adjustable in width from 3 to 6 inches and should be closed at night.

HOUSING TO BUY

The illustrations show the variety of designs available commercially. Since backyard poultry-keeping is less popular than it was thirty years ago, you can still occasionally find such houses second hand by advertising locally - but look carefully at the extent of the almost inevitable wood rot.

These illustrations are all from the latest catalogue of Park Lines and Co., one of the few remaining suppliers of small-scale poultry housing. Their address is:

Park Lines and Co.,
Park House,
501 Green Lanes,
London N 13 4BS.

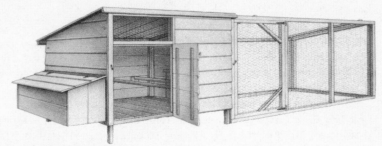

"BELL" HOUSE AND RUN

"MIDWAY" for 10—12 birds.

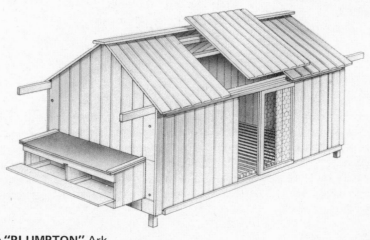

The **"PLUMPTON"** Ark

51

DO IT YOURSELF

If you have a saw, a hammer and a little confidence at D-I-Y,
you can save quite a bit on housing. Either follow the traditional
designs or make your housing smaller, lighter and less sophisticated, as I propose.

My housing design is based on a standard cross-section, used
in a length to suit the number of hens to be housed. I have designed
it to be as simple and light as possible, consistent with the basic
requirements described above. It is also designed to accomodate
growers and to be used with covered runs at either end.

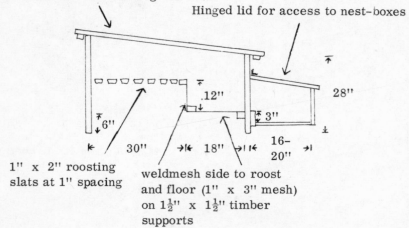

Lid lifts off for cleaning out

Hinged lid for access to nest-boxes

.12"

28"

6"

3"

30"   18"   16-
                20"

1" x 2" roosting
slats at 1" spacing

weldmesh side to roost
and floor (1" x 3" mesh)
on 1½" x 1½" timber
supports

The framing is 1½" x 1½" timber. The cladding can either be
¼" exterior ply for lightness and cheapness or ½" tongued and
grooved boards for better looks. The internal length of the house
depends upon how many 20 inch nest boxes are to be accomodated.
For up to 4 layers, I suggest 2 boxes; up to 8 layers, 3 boxes;
up to 12, 4 etc.. That leaves space for expansion and for growers
to roost as well.

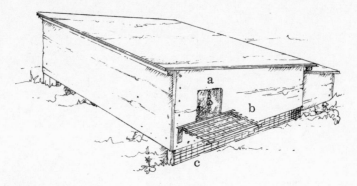

a) anti-fox chain
b) foot-scraper entrance at 18"
c) weldmesh end

Two 4 feet wide by, say 28 inch high runs can then be constructed. One for the growers at the far end and one facing the main entrance for early mornings. Each is made with one open end to butt against the henhouse end wall.

In addition, a small broody box should be constructed, which measures 20 inches square and which doubles as a box for curing broodies and for housing a mother hen and her chicks.

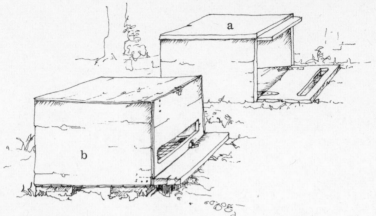

a) As nest-box for raising chicks, with a separate roof covering the weldmesh face and with the bottom-hinged door open.
b) As broody-cure, with weldmesh face as floor, raised on bricks and with door closed. The 3 inch hole in the door allows the broody to reach food and water placed outside on the shelf 3 inches below.

Housing for deep litter follows the same basic principles regarding roosts and nesting boxes, but must also provide for feeding, watering and daytime exercise space. Because everything is inside, the roosts should be placed over a droppings pit, inaccessible to the hens but accessible to you for cleaning out. The only other difference is the need for plenty of ventilation.

For feeding inside, a self-feeding hopper is used, hanging at just below beak height to reduce wastage. I do not know whether anyone has tried scattering feed in the litter, but it could be a way of reducing boredom. The Ministry of Agriculture Code for Domestic Fowls suggests a maximum stocking rate in deep litter of 3 lb live weight per square foot, including roosts and nesting boxes. This is very close packed and something more like 3 - 4 square feet per $3\frac{1}{2}$ - 4 lb bird would be preferable. To give them maximum exercise space, nest boxes can be placed about 18 inches off the litter so that there is more room to run around below them. If you do this, give the birds a landing platform just in front of the nest boxes and remember to place the roosts above the nest boxes, as before.

In deep litter houses, nest boxes are generally placed completely inside the house. So it is necessary to discourage the birds from perching on top of them. This can be done with wire netting or by giving the nest boxes sloping tops at an angle of 45° or more.

second stage: your chicks arrive

I assume that you are now ready to take the plunge. Hopefully, you will. having read this far, feel a little more knowledgeable about keeping chickens. You will have decided on the system that suits you best; on how many hens to keep, and of what breed or make. You will already have tracked down sources of bought-in feed and have bought or built the henhouse, runs and fencing.

You are now just about ready to order your chicks from the hatchery. But first - when should you do it and how should you rear them?

7 rearing chicks into layers

TIMING

In Chapter Two, I described the natural annual egg production
cycle and the fact that growers are encouraged to come into lay
by lengthening days and discouraged by shortening days. As to
timing your start in poultry keeping, you have two choices. You
can get them hatched in late February to early April, in which
case they should come into lay around August and September
and give you a modest supply through the Autumn, when eggs
are scarcest and at highest prices. Alternatively, leave your
start until August, in which case they will come into lay with
the lengthening days of the following January.

For starting off, I recommend August-hatched chicks. Foster-
mother broodies are at their most plentiful in July and August
and virtually unobtainable between February and April. Also, you
will be faced with kinder weather and fewer discomforts by starting in August. However, if you are really impatient, any time between April and October will do.

REARING

In the bird world, there are basically two types of chick. The
'altricial' chick is born totally helpless, featherless, blind and
unable to move from the nest. He has one main instinct: to open
his mouth wide and screech at his parents to bring food. The
'precocial' chick is born covered in downy feathers, able to walk

in minutes and capable of finding his own food within two or three days. Luckily for you, the chicks of all domestic poultry are precocial.

That is why they can be obtained from hatcheries as 'day-olds' - they will survive without food for another day or two - although the quicker they are in to permanent quarters, the better, of course. However they are not born with a 100 per cent efficient central heating system. This takes them a few weeks to develop which means that, for their first few weeks, you have to arrange to supply outside heat. Nor are they born with a 100 per cent efficient digestive system - they can handle greenstuffs, small insects and seeds, but bigger things like whole grains have to be cut up for them.

Hence the special needs you have to cater for when rearing chicks are:

- warmth
- food chopped into manageable bits
- as with all young animals, food rich in the body-building nutrients: protein and calcium

First, the warmth. In Nature, this is supplied by the mother hen. The chicks snuggle next to her and under her wings to get the warmth they need. When rearing day-olds, you have the choice of using a natural foster-mother (a broody hen), or using some sort of artificial brooder to provide the warmth.

You will not, by now, be surprised to learn that I prefer natural brooding. I know in my bones that it is better for a chick to feel the real warmth, heartbeat, feathers, movement and re-assuring clucks of a mother hen than the sterile heat of an artificial brooder. Not only that, but the mother hen never lets them out if it is too cold or if other dangers lurk; she pecks any big food lumps into small pieces for them and best of all for the backyarder, teaches them from an early age the techniques of searching for wild foods.

Having said that, natural brooding does present difficulties when you are just starting. Where do you get the foster-mother? Well, there is a modest market in broody hens and by scouting around local backyarders, gamekeepers or even the livestock small ads of your local paper, you may be able to buy or hire a broody for the necessary 6 weeks (or 8 in colder weather). There is a danger that she may bring disease or parasites in and give you a bad start, so be aware of that risk. I am a risk taker and I feel strongly about natural rearing, but I can quite understand

others opting for artificial brooding when they make their start.
I have given you the pro's and con's; the choice is yours. Here
are the two methods.

NATURAL BROODING

When you locate your broody for sale, hire or loan, make sure
that she is from a healthy-looking flock kept in clean conditions.
Agree the deal on condition she turns out actually to be broody,
but you can check for this by feeling under the hen's breast - it
should be very hot - like a severe fever. A purple tinge to the
comb is another good sign.

When you get her home, put her in the broody box, in which
you have arranged a small sitting of eggs, either real ones or
china ones, in some dry litter. Keep her out of direct winds and
close up the box at night. Provide her with food and water just
outside, preferably ad lib. Also provide a dust bath of ash or
fine dust, to which has been added some Derris powder. This
is a natural insecticide which will help to free her of any parasites she has brought with her.

If, after at least 4 days, she is still patiently sitting on the
eggs, you can pronounce her well and truly broody. Go ahead
and order your day-olds for delivery as soon as possible, but
certainly within another fortnight. When ordering, remember
that you may get losses on the way. If you want to end up with
an average of 7 layers, order 9 chicks. Up to 4, order an extra
1 chick; above that an extra 2 chicks.

Introduce the chicks at dusk. Do it very gently and quietly,
talking reassuringly to the hen as you do so. She will probably
peck your hand protectively when you reach under her, so wear
a leather glove. Reach in and take out one egg, palm down so
that she cannot see. Now carefully replace it with chick No.1
and leave them both alone for 20 minutes or so. If the chick has
not been rejected by the time that you get back, all is well. Take
out another egg and put in chick No.2. Leave them for 10 minutes
or more and repeat this until all the eggs are out and the chicks
in.

If you are unlucky and she rejects No.1, leave her for an
hour to forget about it and try the operation again with a different chick as No.1. If it becomes clear that she steadfastly re-
fuses to adopt your orphan chicks, you had better hurry and rig
up a simple artificial brooder as quickly as possible.

But let us assume that you are lucky and that she accepts

them. Using the broody box and small chick-proof, predator-proof run as described in the last chapter, the foster-mother will now look after them. For the first week, she will venture out very little except in really warm weather. After six weeks (eight in colder weather), the chicks no longer need outside heat and the foster-mother can be moved back into the main run - but this will only work if she is kept apart from her ex-brood by keeping them to the rearing box and run.

Until they are twelve weeks, the growing chicks can stay in these quarters. After twelve weeks, they are capable of perching and should be introduced to their eventual adult quarters. In my design of housing, this can be done by simply opening the small adjustable growers' door to give them access from their run to the henhouse, at the same time removing their rearing box. This should be done at night and the growers placed by hand onto perches.

Check that they are perching properly every night for the first few nights. This is the best time to train them out of the bad habit of sleeping in the nest boxes. This is something you should do, by the way, with any bird introduced into your hen house, of whatever age.

For later on, when you are introducing replacement growers to established layers, some authorities suggest fitting out your hen house with a wire netting partition, so that the older birds can see the growers and get used to their presence, but cannot get at them. If you have trouble getting the growers to perch, try this. It may help to make them feel more secure.

ARTIFICIAL BROODING

The simplest artificial brooder is the hay box brooder - just a very well insulated box which keeps the chicks warm without fuel or electricity.

The floor is best made of $\frac{1}{2}$" mesh, with the box placed over a layer of dry litter. This can then be changed regularly and the box kept clean fairly easily. The floor mesh should be stretched tight on the box frame. A wooden outer lid is needed, but at a height of about 4 - 6 inches over the floor, an inner nest top of $\frac{1}{2}$" mesh should be placed over the nest space and covered with loose straw. The effect is a cosy space surrounded on all sides by insulation. The nest space itself is best round without corners. Chicks do tend to huddle in corners and the first one in can get suffocated.

As to the actual size of the nest, this is a question of trial and error. If, when you open up for a peep, the chicks are all huddled together, they are not getting enough warmth. That means looser hay, more of it or a smaller nest area will be required. If, on the other hand, they are all to be found round the edge of the nest, against the wire, then it is too hot and the reverse action must be taken. A rule of thumb is 5 day-olds per square foot of floor area.

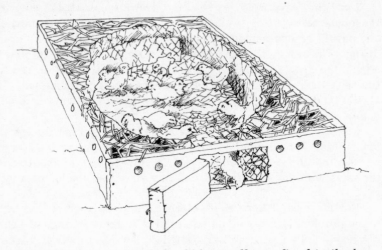

For the first week, they should be totally confined to the hay-box. That means there will have to be enough room inside for a small container of food and another of water. Try to arrange these so that they do not tip up. One way is to wire the containers to the mesh floor. The box is best kept inside, say on the kitchen floor, during this first week.

After that, the hay-box can be taken out to the chick run and the door opened once the morning has warmed up. Their feed can be scattered and the drinker moved outside. Gradually increase the nest area and reduce the insulation as the chicks grow, but check them at night for huddling. At 8 weeks of age you should be able to remove all the insulation and leave them with just the litter underneath. After that, the story is the same as for natural rearing.

There are lots of other designs of artificial brooder for small scale use. Most use an ordinary 100 Watt bulb 15 inches above the floor as the heat source. This is raised 2 inches higher each week. I cannot believe that such fierce light is good for sensitive little chicks, so I don't go for these brooders much. Also, why pay good money for electricity, when hay works as well? If you are keen, you can find out more in any of the older poultry books listed in the back, however.

FEEDING

To recap, the two basic feed requirements on which young stock differ from adults are firstly that they need more protein and calcium and secondly that grains must be fed broken up. You can buy specially formulated chick rations and even special growers' mash for when they are a bit older, but if they are with a broody and have room to search for wild foods, I prefer sticking to grains, animal meal and green shoots. We grind the grain (wheat, barley, maize or whatever) to the size of the large modern instant coffee grains. If you do not have a grain mill, you will have to buy a small quantity of 'kibbled' or crushed grains from the local pet store or corn merchant.

After about the fourth week, the grain can be fed whole. Actually, using a broody, you could probably feed it whole right from the start, as she would no doubt break it up for them. I do not know anyone who has tried this, but I will try it myself next Spring.

The protein and calcium 'extras' are in the form of extra rations of bone-and-blood or fishmeal. Instead of the adult ration of about 10 parts grain to 1 part meal, I reckon that 5 to 1 is more like it for chicks. Like everything else, this depends on how much they can forage for themselves. Since growers are usually reared in Summer, they, like our layers, are expected to find their own extra protein and calcium once they are allowed out on range. Their grain is also reduced proportionately, as is the layers'.

As to quantities, that is very difficult. Obviously the mother hen needs her summer ration, out of lay, of about $2\frac{1}{2}$ to 3 oz of grain. In addition, the chicks need a gradually increasing ration of the grain and meal mixture. They also need to eat little and often, so for the first week in an artificial brooder, be on the safe side and leave it ad lib. Once they feed outside, sprinkle food around the run and that should keep them eating most of the warm part of the day.

Here is a useful rule-of-thumb for chick feeding. It is not accurate, but the general health and shape of your chicks should tell you if it needs adjusting.

For every 4 chicks, feed 1 oz of food per day between them

for the first week. In the second week raise the quantity to 2 oz between them, and in the third week to 3 oz. Carry on increasing the diet at this rate until the tenth week when you will be feeding 10 oz per day between them. Thereafter, increase the ration of daily food by a further $\frac{1}{2}$ oz until they are twenty-two weeks old and eating 4 oz of food per day each. (Remember, if you are using a broody, to add on her $2\frac{1}{2}$ to 3oz, since she cannot be fed separately.)

MARKING

You may like to monitor the individual progess of your new chicks - for instance, their weights at different ages, their behaviour, activity etc. If you do, the method of marking chicks is to nick the webs between their toes to a depth of one eighth of an inch. It is claimed that if this is done quickly with a scalpel or a very sharp knife, it is harmless and painless. Once the birds are at Point of Lay, you can fit them with leg rings.

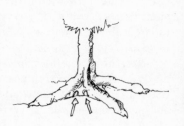

A NOTE ON HATCHING

So far, I have assumed that you are following my recommendation to buy in day-olds when you want to rear replacement laying stock, but you may prefer to hatch your own, or at least hatch a few birds for meat-eating (see also Chapter Nine on meat).

Hatching your own stock should only be attempted under a broody - artificial hatching is a tricky affair. Nest the broody on a shallow bed of small-particle dry litter. Some of the damp of the earth should come through. Choose perfectly shaped, blemish-free, medium-sized eggs as fresh as possible. Ten to twelve is a reasonable clutch, but a large hen can handle up to twenty in Summer. The broody may need to be persuaded off them to feed and back on again for the first few days. In twenty-one days from the date she starts warming them, you should see chicks appear.

If some eggs still remain unhatched after twenty-one days,

test them by placing gently in a bowl of hot water. Those that
sink, then bob about after a few seconds contain live chicks
and should be put back under the hen until they hatch. If, after
ten minutes in the water there is no sign of life, the chick is
'dead-in-shell'.

Some older books advocate all sorts of 'midwifery' to help
out chicks in difficulty. So long as the nest has not been allowed
to get too dry, then my feeling is that all chicks healthy enough
to be worth raising will hatch naturally. Why make trouble for
yourself later by helping out the duds?

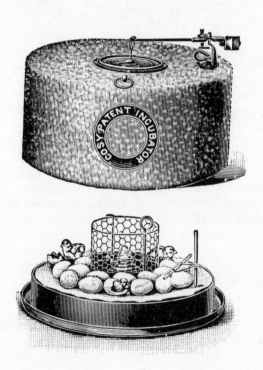

8 when to re-stock

You may have assumed that having a few hens is a bit like having a cat or dog in that you keep them until they get old and smelly, or until they die. Well, it is not like that. If your hens lay 240 eggs in their first year, they will only lay 80 per cent of that number, around 190, in their second year, then 80 per cent of that figure, 150, in their third year, and so on. But their feed consumption stays about the same.

The cost of buying a chick and of the feed needed to get it to laying age is 80p - £1. The cost of feeding a layer for a year is about £2. In the first year she will lay you about £7 worth of eggs at retail prices. In the second year, she will lay about £5.50 worth. In the third, about £4.50. She will go on making a 'profit over feed costs' until about her sixth year, but you can see from the figures that you are better off if you kill hens after their first laying year and rear up replacements every year. That, in fact, is what nearly all commercial poultrymen do.

But there is more to it than pure economics, at least there is for me. At the end of her first laying season a hen has just about reached full size. It seems pretty young to die. I prefer to keep them for two seasons. The return is slightly lower but I feel better about it. There are two modest compensations for this policy: you get an increased proportion of large eggs in the second season and the hen, when you kill her, gives you a good $\frac{1}{2}$ lb more meat.

Based on a two season lifespan, you can either rear replacements for half the stock each year, with the result that your egg production level stays stable from year to year, or you can replace all the hens every two years, which is slightly less work. My family are so fond of rearing chicks that it gets done every year, but it does not have to be that way.

Under natural conditions, a first egg-laying season starts, depending upon the date of hatch, some time between August and March. In all cases, it ends with the moult in September or October. The second season starts between November and January and goes through to the following September or October, when the hen starts her second moult and should be killed. I suggested that a beginner should start with August-hatched chicks, in which case he gets a relatively short first season. When rearing new stock to replace those to be killed in September or October, however, I advocate getting chicks hatched in early Spring. This ensures a reasonably steady supply of eggs and also gives the second and subsequent batches of chicks longer first laying seasons.

If you go for artificial brooding, get chicks hatched before mid-May - the earlier the better. If you are like me, a natural brooding fan, then you are dependent upon one of your hens going broody. This can be encouraged to some degree by removing the partitions between nest boxes so they can see each other laying and also by packing the boxes with much more straw so that they get much warmer when they sit on the nests. It also helps to leave as many eggs as you can afford in each nest (or buy Eltex china eggs). If you do all this in late February or early March, it may help you to get a broody. Once you have one, revert fast to normal conditions.

Having said that, it must be admitted that modern hybrids are very slow to go broody. I do not really know what the answer to this is. Do we give up natural brooding or develop better methods of inciting broodiness (artificially heated nest boxes with mirror side walls, perhaps)? Another answer is to keep one bantam hen. They are notoriously good broodies (and consequently poor layers) and they eat very little. The extra £1 per year in food costs could be justified if she rears 10 or 12 hybrid layers early enough to give you good winter laying. See Chapter Fourteen for more on Bantams.

To try to get all these dates a little clearer, here is a suggested time plan for a family of four, requiring an average of 7 hens, restocking every other year:

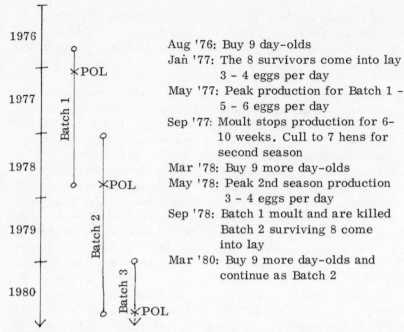

Aug '76: Buy 9 day-olds
Jan '77: The 8 survivors come into lay
3 - 4 eggs per day
May '77: Peak production for Batch 1 -
5 - 6 eggs per day
Sep '77: Moult stops production for 6-
10 weeks. Cull to 7 hens for
second season
Mar '78: Buy 9 more day-olds
May '78: Peak 2nd season production
3 - 4 eggs per day
Sep '78: Batch 1 moult and are killed
Batch 2 surviving 8 come
into lay
Mar '80: Buy 9 more day-olds and
continue as Batch 2

Note that 9 day-olds are started, with the expectation that 8 will
survive to Point-of-Lay (POL on chart). At the end of the first
season, only the best 7 are kept on to do a second season. Picking out the least promising is called 'culling'.

CULLING

There are two times to cull. Firstly, when you notice signs of
obvious weakness which cannot be cleared up. The hen in question is culled as uneconomic to keep on. She is killed and, if not
diseased, eaten. The other time is at the moult, when assessing
which of the batch is the least promising. This is a way of boost-
ing the average performance in the second season.

During the season, test your hens from time to time as to
whether they are in lay. I do this at night, when they are easier
to catch and handle. Hold the hen as in the picture and feel for
the edges of the two pelvic bones one each side of the vent. If
they are at least two finger widths apart, this is a fair indication
that she is in lay. Feel now for the end of the breast bone further
under the hen, between the legs. If you can get four finger widths
between the breast bone and the pelvic bones, that is a further sign
that she is in lay.

The former test is the better. If still in doubt as to the hen being in lay, isolate her (I put her in the broody box and run) for a week with plenty of food and water and preferably within sight of the rest of the flock. If there are no eggs after a week, cull her.

Culling at the time of the first moult is more tricky, since we assume that they are all out of lay temporarily. You have to go on other signs of general health and laying potential:

Early or long moult
- The early moulters are suspect as are the ones that take longest to complete the moult, although this is more difficult to observe. These are both signs of lower laying potential and said by some to be the best signs.

Weight
- Weigh all hens. The heaviest should be examined for signs of going to fat. Pinch the skin near the breast bone to feel if there is much of a layer of fat beneath it. The lightest should also be looked at for signs of general low health - their lightness may be a sign of lack of vigour in collecting food.

Fullness of crop
> - Test this a couple of hours after roosting. The emptiest crops are suspect as poor food collectors.

Other signs of poor food gatherers
> - Long nails mean they have not been scratching much. A long curving beak also suggests under-use. Any lack of keenness and speed when you come to feed them is a very bad sign.

Other signs of general ill health
> - During the moult, these are very difficult to assess. It is as well to make a first assessment in, say, July, when they are still in lay and make the final decision at the moult. A bird in lay should have well-developed comb and wattles (the red bits that hang down behind the beak); a bold eye with no sign of sunkenness or dullness and should have tight feathering. In July, the feathers of a good layer around her vent should look pretty worn and scraggy after a full season of heavy laying.

If a large proportion of your stock show signs of ill health, something is seriously wrong with your management. Otherwise just pick out the one that scores clearly lowest on the above list of points and kill her for eating. In a bigger flock, I would cull 10 - 15 per cent at the moult, depending upon how clearly substandard the birds in question were.

FORCING THE MOULT

This is a technique for getting the moult over early in the Autumn so that the hens definitely come back into lay fast and provide eggs when they are at their scarcest, in late Autumn. In my view, it is applicable to commercial production but not to backyarding. I prefer to store some eggs from Summer in waterglass (see Chapter Nine) and let the hens moult naturally. So long as I have culled well, I feel confident that this will eventually pay off, even if we are low on fresh eggs for a few weeks.

The technique is not universally successful. It consists of feeding only bran and water until the hens are clearly moulting, but for at least ten days. Keep them off range, of course, otherwise your efforts will be wasted.

SPEEDING THE MOULT

The quicker the birds can grow their new feathers and have the rest they need, the quicker they will be back in production. For feather production, feed oil seeds (rapeseed, linseed, sunflower etc.) and double rations of animal meal.

9 the harvest

After eight solid chapters of what to do and what not to do, at last the payoff... Oh, the joy of collecting warm brown eggs from just up the garden... Even after years of it, we still get pleasure from the feel and shape of them. When you keep a few hens, you can even recognise the shapes, spot patterns and the shades of particular hens. You feed and shelter them, but in return, your hens present you with such a fine harvest. No need to milk it out of them or kill them for it - beautifully packaged portions of near-perfect food are presented to you daily on beds of straw.

Eggs are to me the justification for keeping hens. The meat, feathers, manure and improved land are extra bonuses. Here, in this short chapter, are some collected hints on utilising all these products.

EGGS

Collecting

The more you take away, the more they lay and the less they go broody. Visits at 10, 2 and 5 o'clock would be ideal, but this is not critical. If you use a drop-through nest, you can leave the eggs as long as you like. Hens on free range may find alternative nests out in the fields or hedgerows. Listen out when you hear laying noises coming from new places. Track them down, collect the eggs and confine the birds for a few days. Also, try to make such places less attractive.

Storing

Eggs will keep at room temperature for about three weeks, but
I prefer mine eaten within 3 days. It can be tricky in Summer
keeping track of which eggs in the rack to use next. Here is a
solution. Suppose you consume, or wish to consume, 4 eggs per
day and also decide to keep a maximum stock of 12 eggs at any
time. Get a flat egg-tray (from a market stall or shop that still
sells them in that way) and mark it up as shown.

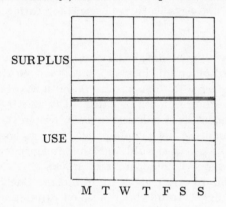

Store eggs in the row corresponding to the day of the week
laid. Fill the first 4 places first. Any laid above 4 go into the
corresponding "SURPLUS" slots. If there are already 12 eggs
in the "USE" rows, put the new extras into "SURPLUS". Use
the oldest eggs first. Once you have 6 "SURPLUS" eggs, sell
them. They will still be pretty fresh using this method.

Alternatively if you do not wish to sell the surplus eggs, put
them straight into the preserving bucket for use in Winter.

Selling

So long as you sell eggs as a producer <u>direct</u> to consumers and
not via an intermediary, you are under no legal obligation to
grade them, now will you be under the Common Market regulations, as I understand.

The best way of selling an irregular supply of small surpluses,
frankly, is to friends and neighbours. It is only worth putting a
notice at the gate if you keep enough hens to give you a regular
surplus. If your eggs are big, brown, fresh and rich-tasting, you
can soon build up a regular clientele locally. If there is someone

at the house all day and if you have the land, this can be quite a good source of extra income, particularly for the house-bound wife with young children.

Eggs for sale should be clean. Any damp on the shell reduces the egg's storing life, so clean with steel wool. Only use a damp rag as a last resort. The best way to produce clean eggs is to keep the nest boxes clean.

If you keep a cockerel running with your hens there is also a possibility that you might be able to sell fertilised eggs to individuals who have a broody but no cockerel. The asking price is greater than if the eggs were being sold for eating.

Preserving

Eggs are preserved by making their shells impervious. They will keep in a cool place for about six months. The simplest method of sealing eggs is to submerge them in waterglass. This is a solution of sodium silicate in water. It is usually bought in the form of a concentrated solution in a tin, which you dilute as per the instructions. So few people preserve eggs now that it is difficult to get hold of but try old fashioned independent chemists and ironmongers. Any ironmonger who stocks MANGERS products should be able to order you a tin of theirs. One 250 ml tin will preserve 80 eggs. Waterglass is also available by post from Self-sufficiency and Smallholding Supplies, The Old Palace, Priory Road, Wells, Somerset, who also supply Eltex equipment and Park Lines poultry houses.

Use an enamel bucket with a lid. These are expensive but do last for ever. Waterglass attacks glass and if you lift a plastic bucket full of eggs, half will crack. Use only same-day eggs to put down and make sure that they are clean and free from cracks. Preserved eggs can be used just like fresh ones, except that if you boil them, prick a pin-hole in the blunt end first to stop them splitting open with the pressure of trapped air. Naturally, they do not taste as good as fresh eggs, so we use them only for cooking, keeping the limited Winter supply of fresh eggs for eating boiled or fried.

As an alternative, eggs can be deep-frozen to keep for up to 10 months. Either freeze yolks and whites separately in small quantities or freeze them whisked lightly. They can be thawed quickly and used immediately.

Just in case waterglass becomes unavailable, you can preserve in lime water. Add 4 parts of freshly slaked lime to 20 parts of water. Stir well each day for 5 days. Add 1 part of salt.

Stir daily for 3 days. Strain and use as waterglass, but do not break the crust that forms, it reduces evaporation.

Using eggs

A recent survey (Harper Adams West Midlands Survey, 1973) carried out in the Midlands showed how unenterprising people tend to be in their use of eggs. 45 per cent were fried, 29 per cent boiled, 11 per cent poached, 8 per cent scrambled and only 7 per cent used in other ways. 47 per cent were eaten at breakfast, 30 per cent at tea and only 18 per cent at lunch or supper. If you keep hens and carry on like that, you will soon get very sick of eggs. In Appendix III, you will find a checklist of ideas for other uses of eggs. This is not a recipe book, but most of the ideas listed can be found in any general cookery book.

In addition, the British Egg Information Service, Haymarket House, Oxendon Street, London SW1 produce a series of free recipes and leaflets.

When using eggs, it is as well to be aware that some authorities advise against eating egg white uncooked. Apparantly, raw white passes through the stomach so quickly that very little of the nutrients reach the blood. The white should be cooked until beyond the semi-liquid phase and at least semi-solid. Also be aware that Vitamin B2 is destroyed by light. Even during frying in an open pan, an egg can lose half its B2 content. So, if you value your vitamins, keep the lid on when frying or scrambling eggs.

POULTRY MEAT

Killing

A bird to be killed for meat is better for being kept separately in a small confined run for the last two weeks or longer, and being fed on barley and water - as much as it will eat. It should be given no food for its last 24 hours, just water.

The best way to kill a chicken is to dislocate its spinal cord at the top of the neck. Whichever way you do it, killing a hen which you have kept and has laid well for you is an unpleasant job, but at least with dislocation, there is no blood and she feels nothing.

Hold her legs in the left hand, stand up and grip her behind the head with the right, placing the index and middle finger on either side of the head and the thumb on the top of the head. Extend the neck gently as far as it will easily go. Prepare your-

self for the kill. Give a sharp firm pull to extend the neck 9 inches more, giving the head an upward twist at the same time by means of a movement at your wrist. You should easily feel the neck dislocate cleanly. Pull too hard and the head will come off and you will make a bloody mess. Pull too little and the poor hen will still be alive and <u>very</u> perturbed. If you brace your muscles for a firm 9 inch pull, you should be alright.

The spinal cord dislocates just behind the head. The bird dies instantly and if hung upside down, the blood drains conveniently into the cavity formed in the neck. Of course, the wings will carry on flapping for a couple of minutes after death, but you can check the bird is 100 per cent dead if the head falls completely loosely with clearly no connection left at the spinal cord.

The best plan is to tie its legs with twine before killing. Once the neck is broken, hang it up from a hook somewhere to finish flapping.

<u>Plucking</u>

The best time to pluck is as soon as the bird stops flapping. Do not wait. Sit down with an apron over your legs and two containers on the floor for the feathers. The head of the bird should hang downwards throughout to facilitate the draining of the blood. The long wing and tail feathers are the toughest and should therefore be done first. Pull them individually with a short, sharp action. Start the smaller body feathers at the breast, pulling about 3 at a time with sharp pulls. Put the long feathers in one container and those suitable for stuffing cushions etc. in the other.

After plucking all over up to within 3 inches of the head, you will notice that a few feather stubs have inevitably been left behind. Pull them out between a sort, blunt knife and your thumb (called 'stubbing'). If there are also hairs left on the skin, singe them off over a gas flame or meths burner. Finally, squeeze the vent to get rid of excess excreta and wipe clean. Hang the bird in the cool until you wish to prepare it for cooking. Do not let it hang too long and watch for greening around the vent.

Preparing for Cooking (or 'trussing')

Like all these rather unpleasant operations, the best way to learn is by watching someone who knows how. It should not be too difficult to arrange to watch a local butcher or poultry-stall man do the job. In this section, I will at least attempt to show you the stages to look out for.

1. Cut the skin around the shank just up from the foot joint. Then break the leg at this point over the edge of the table and pull hard on the foot until the sinew pulls out suddenly and you remove foot and sinew at one go. It is not necessary to remove the sinews from very young chickens.

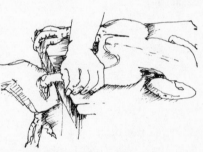

2. Put the bird breast-down on a chopping-board. With a sharp 3-4" bladed kitchen knife, cut half-way round the neck-skin at a point 1" up the back of the neck from the shoulder.

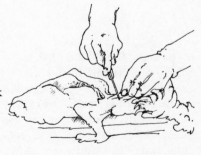

3. Slit up from this start 3" up the neck and cut off the resulting flap of neck-skin. Twist the neck out of the body or cut it at the lowest joint.

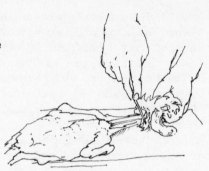

4. Work a finger round the crop and cut it out as far down as possible.

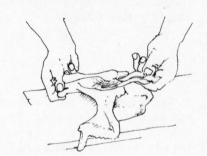

5. Turn the bird onto its back and poke your finger into the neck-hole. Loosen the lungs and other organs from the rib-cage.

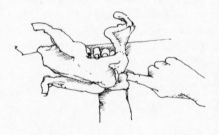

6. Balance the bird on the neck-hole, hold the tail in the left hand and make a 3" cut between the tail and the vent. Feel into this cut with your finger for the intestine. Holding it clear, cut right round the vent.

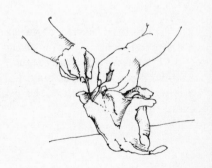

7. Enlarge this hole up to the breast bone until you can get a thumb and two fingers inside the bird. Feel for the hard, golfball sized gizzard and pull it out. With it should come the liver, heart, lungs and intestines. If they do not move then you have failed to dislodge them properly at Step 5.

8. Either joint the bird or tie it up ready for the oven. The purpose of tying a bird for sale is to make the breast look as big as possible, by tying the wings in low and the legs back tight. For this you will need a trussing needle, but frankly, I feel it is adequate for home use to tie the legs down to the parson's nose (the tail).

The 'giblets' consist of neck, heart, liver and gizzard (split lengthways and the inner horny skin containing the stones and food removed).

Rearing up extra stock for meat

If you follow the plan suggested in this book, the only birds you will be killing to eat will be your old layers. These will supply you only one bird per family member per year. If you like more poultry meat than that, I suggest that you rear young cockerels. Day-old cockerels of the Light Sussex/Rhode Island dual-purpose cross can be bought fairly cheaply, since they are really just 'by-products'. They rear up into very acceptable table birds. Do not buy day-old cockerels of a hybrid laying strain - there is no flesh on them. The alternatives are to buy day-old broiler hybrid chicks or hatch your own from a meat-based cockerel kept as sire to the laying flock. See Chapter Seven on Hatching and Rearing.

The broiler hybrids will fatten up to 4 lb liveweight in an incredible 8 weeks, but will not, of course, taste of much. Your own or bought-in cross cockerels can be fattened up much longer

- even to 7 months of age to a weight of 9 - 12 lb if you so wish.

Rearing is as for bought-in laying stock, except that no animal meal need be fed for birds with access to range. Oats and barley produce fine table birds. If you want your cockerels to fatten up fast and to produce tender meat (say to 8 lb liveweight at 20 weeks), one way is to caponise them. This is a sort of chemical castration carried out at 12 weeks. Boots Farm sell a kit for 100 birds at £1.40 with full instructions. Basically, the job consists of implanting a tiny hormone tablet under the loose skin at the back of the neck, using an applicator supplied in the kit.

Please be aware, however, that certain authorities disapprove of the practice of adding hormones to animals that will later be eaten, anxious that these hormones might have the same fattening, softening effects on the people eating them as they do on the cockerels....

FEATHERS

The quantity of feathers obtained from backyarding is hardly worth the bother of degreasing them to preserve them, but if you are an enthusiast, here is the traditional method.

Make lime water by adding 1 lb of quicklime to a gallon of water, stir it well and let it settle. Syphon off the solution, leaving the sediment. Steep the feathers in the solution, kneading them to get all the grease free. Leave them to soak for 3 days. Drain off any floating impurities, remove the feathers and wash them several times in water. Dry them, using a low oven, and store in paper bags. Watch for moths.

Feathers also have a use in the garden as a source of slowly released nitrogen beneath soft fruit.

MANURE

A hen produces 1 - 2 hundredweight of high-quality manure per year. Its composition is, approximately,

	%
moisture	55
organic matter	29
nitrogen	2
phosphates	1.2
potash	0.6

Compared to farmyard manure, it is rich in nitrogen and

phosphate but deficient in potash. It is rather strong to apply fresh.

In Chapter Three, I explained how the strawyard system can be used to fertilise land directly, without the need to move muck around. Nevertheless, you can, if you prefer, use droppings and spent litter, kept one year in a compost heap, to great effect. Its composition is approximately:

	%
nitrogen	2.2
phosphates	2.5
potash	1.2

If kept two years, phosphate and potash yields improve further to 3.4 and 1.8 per cent respectively. After two years the nitrogen yield falls and no further improvements take place. Hence the old farming idea that poultry manure on land works well for two years, giving nitrogen in the first year and phosphate in the second.

To make a usable concentrated fertiliser from the droppings pit, add in 10 per cent of superphosphate (1 oz/bird/week) and keep it for one month, by which time it has broken down to a dry, friable, odourless fertiliser. Alternatively, put 2 lb of sedgepeat per bird in the droppings pit. Remove peat and droppings and store for three months.

USING POULTRY TO IMPROVE LAND

I dealt in Chapter Two with the adverse effects poultry can have on land, using a dirt run or unrested grassland range, and also of the beneficial effects on land of the strawyard system.

The strawyard system can be used in order to clear and improve new land for crops. Hens kept intensively directly on grass will do a thorough job of reducing grass cover and clearing nematodes, insects and weeds. But when using them in this way, it is as well to balance out the acidic effect of their droppings by liming the soil well before planting.

A novel method of using hens to improve land has been developed by Bernard Capon, an organic farmer. He feeds a small proportion of whole grain into the diet of his cattle and then lets his hens onto the land just vacated by the cows. The hens scratch the dry cow pats for grain, at the same time doing the very useful job of eating harmful nematodes in the pats and spreading the manure more evenly over the pasture. The hens are stocked at the rate of 6 to the acre and operate as very low-cost harrowers and pesticides, with a few eggs as by-products. This must be one of the most ingenious uses of the domestic fowl yet devised.

10 diseases and other problems

In my opinion, unless there is a danger to other stock or to humans, the services of a vet are now uneconomic for the treatment of a small backyard flock of hens. If your stock is from a reputable hatchery; if you keep the birds in reasonably sanitary conditions and if you take the preventive steps I list below, disease should be rare. But suppose a bird does show symptoms of ill-health, the answer is to know enough about the symptoms of the more common diseases to be able to give it some simple, basic treatment. If it does not recover, cull it.

As a poultry keeper, the law of the land makes very few special demands on you but you <u>are</u> obliged to notify the Police or the Ministry of Agriculture if your hens contract Fowl Pest or Fowl Plague. So that is another good reason for understanding a little about symtoms.

But first, the question of prevention. The 'protective' food nutrients: vitamins and minerals, must be fed in adequate quantities. That means whole grain, adequate greenstuffs Summer and Winter plus either live insects and worms or concentrate meal. These foods will help to keep your hens in condition and maintain their defences against diseases.

The most vital measure of prevention, however, is preventing the build-up of parasitic worms, protozoa, bacteria, lice, mites and viruses where the birds congregate. The danger points are the roosts, the ground just outside the henhouse door and the drinker. The roosts, roost-ends and all nooks and crannies of the henhouse should be gone over at least once a year, preferably just before new stock arrives, with a blow-lamp, then the whole house should be creosoted. To avoid build-up around the entrance, make sure that you move the house to a new spot regularly. We do this at least every fortnight. As to the drinker, it must be of a design which stops the hens being able to excrete into the water. Hanging it just above ground level helps to stop them scratching dirt and grass back into it, but in all events, give it a really good swill round with clean water every time you re-fill. Give it a thorough clean in soapy water every month.

It is said that a clove of garlic in the drinking water acts as a good general tonic for hens, keeping up their resistance. It is certainly cheap enough to give it the benefit of the doubt. The moult can be a time of reduced resistance to disease, so it is wise, as a further preventive measure, to shorten it as detailed in Chapter Eight, using oil seeds and extra protein rations.

All these measures are simple and cheap. If you get yourself into the right routine from the start, you are bound to reduce the chances of disease and save yourself lots of problems later on. The checklist of regular routines (Appendix II) is designed to help you in this respect. Tear it out and keep it pinned up near or right on the food container.

Many authorities recommend regular de-worming, every 6 months or so, with a proprietary de-worming drug. My feeling is that this proceedure is an admission of failure as to unhygenic conditions and poor management. The regular de-worming process must, in my view, weaken the birds and upset the balance of helpful stomach organisms.

Notes to Disease Guide

a) <u>Treatment A</u>: Isolate the bird in a small box or cage kept warm and dry. Fast her for 24 hours with access only to water. Then feed a mildly laxative diet (barley with husk, bran etc.) with 1 chopped clove of garlic per day for 10 days. If recovering, continue until fit. If no sign of recovery after 10 days, cull. Move the rest of the flock to new ground.

b) <u>Treatment B</u> (for lice, mites, fleas in feathers): Isolate bird. Feed normally but give access to a dust bath or ash or dust to which Derris powder has been added. Keep this Derris bath for future occasions. Move the rest of the flock to new ground. Blow lamp and creosote house. Return isolated bird to the flock after 7 days if clear of parasites.

c) Most hens have a mild population of worms. Some authorities

GUIDE TO PRINCIPAL DISEASES BY SYMPTOM

The symptoms you are likely to notice first are in each case given in capitals.
The symptoms are not *proof* of the disease named, but indications.

SYMPTOMS

A: CHICKS
1. TOES BLEEDING & obviously being pecked.
2. TOES CURL, walking on hocks at 2-4 weeks old.
3. WEAKNESS IN LEGS, shanks look dried up.
4. FITS, with kicking & wing-flapping, then rapid recovery.
5. CRUSTY SCABS AT MOUTH CORNERS, poor feathering.

6. COUGHING, SNEEZING, shake heads, gasping, gaping.

7. OLDER CHICKS STAGGER & FALL, twisting of head, average 50% mortality.

8. CHICKS STAND ABOUT LOOKING CHILLED, back humped, wings drooped, cheeping when excreting, white diarrhoea, heavy mortality.

B: CHICKS AND GROWERS
9. STAND ABOUT WITH EYES CLOSED, droopy wings & tail, head drawn in to body, blood in droppings.

C: GROWERS AND LAYERS
10. LAYER ALWAYS ON NEST, strains to lay, but can't.

11. YOUNG LAYER SQUATS ON HOCKS, one leg appears cramped, otherwise bird healthy.
12. REDDISH TISSUE PROTRUDES FROM VENT.
13. DISTENDED, DOUGHY-FEELING CROP, convulsive movements of neck.
14. FEATHER PECKING, CANNIBALISM, pecking particularly on backs.

15. BIRDS DEJECTED, REFUSE FOOD, discharge from eyes & nostrils. Sudden high mortality.
16. DROWSINESS, RUFFLED FEATHERS, rapid breathing with rattling sound, green-yellow diarrhoea, nervous twitch of neck.
17. DROOPING WING OR LEG CARRIED. FORWARD & TWISTED.

18. CATARRH, watery discharge from nostrils.
19.   "   "   , plus water-like lesions on comb, wattles, etc.
20. WORMS IN DROPPINGS, loss of condition & appetite, face pale, diarrhoea.
21. SWELLING ON FOOT OR FACE.

Both drug and non-drug treatments are given, where applicable.

DISEASE	TREATMENT
TOE PECKING	Replace litter with cut straw 4" thick to cover toes.
VITAMIN B$_2$ DEFICIENCY	Feed more grass, milk, milk powder.
RICKETS	Feed cod liver oil.
INDIGESTION	No feed for 24 hours, just water. Reduce rations.
PANTOTHENATIC ACID DEFICIENCY	Feed yeast, milk, milk powder.
GAPEWORM (in windpipe)	Move to new ground, treatment A (see below) or treat with BARNITAR powder or similar (if you can down track a stockist).
SALMONELLOSIS	Danger! this may be transmissible to humans. Seek advice from vet or Ministry of Agriculture (MAFF).
B.W.D. (bacillary white diarrhoea, or Pullorum Disease)	I would destroy the lot, since even those which recover will remain carriers. Disinfect housing with formalin.
COCCIDIOSIS	Treatment A or Sulphamezathine.
EGG-BOUND	Heat vent over boiled water. Massage gently with olive oil. Dose with castor oil. Egg should come.
LAYER'S CRAMP	Dose with 1 teaspoonful Epsom Salts in water.
PROLAPSE	Massage with oil & push back, then Treatment A.
SOURCROP	Pour 1 teaspoonful olive oil down throat. Massage crop to get contents moving. Fast 24 hours.
BOREDOM, LACK OF PROTEIN, OR PARASITES IN FEATHERS	Increase protein. Isolate victim & dress wound with Stockholm Tar. Increase opportunities for scratching. If parasites seen, Treatment B. Cull persistent attackers.
FOWL PLAGUE	Notify MAFF. Compulsory slaughter & compensation.
FOWL PEST	Notify MAFF. Compulsory vaccination. Don't delay, it spreads fast.
FOWL PARALYSIS (Lymphomatosis)	Treatment A.
A COLD (roup)	Dose with cod liver oil. Plenty of fresh greenfood.
FOWL POX	Treatment A & dab lesions with iodine. Move all stock to new ground.
WORMS (c)	De-worming drugs, or fast 24 hours & give tobacco-water to drink plus laxative diet for 10 days.
WOUND GONE SEPTIC	Swab with disinfectant till hard & ripe. Lance with sterilized knife & dress with iodine. Isolate bird till healed.

are against giving treatment except in definite cases of loss of condition or egg production.

For the ideas in Treatments A and B, I am indebted to the book "Herbal Handbook for Farm and Stable" by Juliette de Bairacli Levy, published by Faber in 1963. This is a book that stands out, unique in its field.

OTHER PROBLEMS

Egg Eating

The problem of egg eating and how to prevent it has been dealt with already in Chapter Six. If you do get a case of it, collect the eggs more frequently and replace them with china eggs. If you can find out the culprit, isolate her to a spot where you can hear her. Remove her eggs immediately after laying. After 10 days of this, put her back and hope that she has lost the taste for eggs. If not, cull her.

Broken Legs

A hen with a broken leg can be treated with a double-sided splint fixed with sticky plaster onto the leg. She should be kept in confinement until she can walk on it again and the splint can be removed. If it takes too long, the break is obviously severe, and she may have to be culled.

Fishy Eggs

It does sometimes happen that an egg tastes fishy. No-one seems quite clear as to whether this is the direct result of feeding fishmeal or due to some other cause. One theory suggests that if an egg is retained longer than normal in the egg duct, it will taste fishy. If you get the occasional fishy egg, it is hardly worth taking any action, but if it happens regularly, first stop all fishmeal and replace with blood and bone. After that, all you can do is try to isolate the offender and get rid of her. Addition of oil or oil seed to the diet may assist the egg on its way down the duct, but I have no evidence that this actually helps reduce fishy eggs.

Vicious Cockerels

Some breeds of cockerel have a tendency to attack people, particularly young children, without warning. The only solutions to this problem are, either to keep your flock confined and children out, or to make do without a cockerel.

third stage: an introduction to other poultry

As I said right at the start, this book is primarily about keeping the domestic fowl, the good old hen. But there are quite a few other types of poultry you may feel like going on to. I have tried, in the following four brief chapters, to give enough information just for you to weigh up the pro's and con's of these other types of poultry. Beyond that, I refer you to specialised books for if and when you go ahead.

11 ducks

BREEDS

Muscovy ducks are more like geese and will be dealt with in the next chapter. Of the other breeds, two stand out on their own. For egg production, the Khaki Campbell is unequalled. The Aylesbury is similarly the supreme breed for meat.

The Campbell was bred by Mrs Campbell in Gloucestershire out of Mallards, Indian Runners and Rouen ducks. None of these are spectacular layers, but Mrs Campbell must have discovered one of the earliest genetic 'nicks' (see Chapter Four). The Campbell regularly averages over 300 eggs per year.

WHY KEEP DUCKS?

If you like duck meat, Aylesburys are worth keeping, particularly if you have grassland space. They eat a great deal of grass and not very much bought food and will fatten up to splendid table birds in 6 - 8 weeks. A broody hen makes an excellent foster mother and the whole business is very little trouble.

As for Campbells as egg producers, they have many advantages:

- production is high - 300 per year or more can be expected from good stock;
- this laying rate, it is claimed, continues for 3 - 4 years;
- the eggs are bigger;
- Winter production is good, so long as the ducks are

protected from extreme cold;
- ducks are hardier than fowl, need less sophisticated shelter and, when adult, very rarely go ill;
- the Khaki Campbell is a pure breed, enabling the back-yarder to breed from his own stock if he wishes;
- ducks can be used to clean gardens of slugs; clear pastures of liver-fluke and to watertight a pond by their continuous paddling.

They also have some disadvantages:
- there is a prejudice against duck eggs making them less saleable;
- this prejudice is due to food poisoning scares, which in turn are due to the unfortunate fact that ducks' eggs last only one week compared to three for hens' eggs;
- duck eggs can have a strong taste, particularly if fed on wastes;
- the eggs taste rubbery in certain dishes such as omelettes;
- ducks are poor rearers, so you really need a broody hen to do the job;
- someone has to be at home to let them out after 10 every morning.

To sum up, Khaki Campbells are most attractive to those with plenty of grass and a pond who are keen on the 'rich' taste of their eggs and who do not want particularly to sell surplus production.

BASIC NEEDS

Ducks should not be kept in the same area as fowls. Their messy habits increase the dangers of disease in the less hardy fowls. They are fast eaters and will crowd the hens off the food if fed together.

Ducks will happily sleep in the open throughout the year, but it is better to house them with some protection. They lay better in Winter that way, they are safe from foxes and it means that your eggs are easier to find. Ducks usually lay before 10 in the morning, but they do have a nasty tendency of dropping their eggs wherever they happen to be. The answer is to keep them in a house or house-and-run over night and let them out after 10 every morning.

The housing can be as basic as you like, so long as it provides a dry, sheltered space with a thick layer of litter. Separate nest boxes are not required.

FEEDING

They will thrive on plenty of grass. An orchard is ideal. If there is a pond nearby, this will supply the ducks with snails, molluscs, slugs etc. and so reduce their need for supplementary protein.

In such a situation, all they need is a little grain in the morning and again in the evening to get them back into the run. Grain is best fed in a clean trough rather than scattered on the grass. But the morning feed can be scattered on the shallow pond water to encourage pond foraging.

If kept in more confined conditions, you need to add greenstuffs and animal meal to the diet. Ducks will also eat virtually any waste food available, but watch out for off-tastes in the eggs and take care to keep up the greenstuffs and protein.

REARING

Hatching and rearing can be done by a broody duck, but they are not very successful mothers. A broody hen is better. The eggs will hatch in 4 weeks. They should be sprinkled with water every day when she is off the nest to feed.

Ducklings should not be allowed to swim before 4 weeks old. They cast their feathers at about 10 weeks and at that time need special care. Keep them dry and warm with extra protein and oil in their diet.

DISEASES

Adult ducks are much less prone to diseases than fowls. In ducklings, however, watch out for all the fowl chick diseases plus a particularly nasty one, Virus Hepatitis. If a number die suddenly together at under 3 weeks, notify MAFF. This is a notifiable disease and serious, though not common.

SOURCES OF MORE DETAILED INFORMATION

The most comprehensive guide available is "Ducks and Geese", a manual produced by the Ministry of Agriculture and published by HMSO. This is a good publication, but geared to larger-scale commercial duck-keeping. "Practical Duck-Keeping" by L. Bonnett, published by Land Books in 1960 is much more readable and homely, but not as thorough. Alas, both are now out of print and will have to be ordered from your public library.

SOURCES OF DAY-OLD DUCKLINGS

There are a few commercial suppliers of ducklings. To find them, look in the small-ads of 'Poultry World' or 'Farmer's Weekly'. Alternatively, since Khaki Campbells are a pure breed and reasonably uniform in performance, try to track down a local person who keeps them and buy a clutch of fertile eggs for hatching under a broody.

12 geese

As noted in the last chapter, Muscovy ducks, also known as Brazilian Geese, are more like geese than they are like ducks, and will be included in this chapter.

WHY KEEP GEESE?

Geese have two main attractions: their meat and their beligerance. In addition, one can also become very partial to the rich taste and huge size of goose eggs. They are the hardiest and easiest of all poultry to keep, so long as you have the space.

The eating of goose at Christmas is no longer as popular as it was in former times or as it still is in Germany and France. We have tended to go over to the much drier (and frankly, duller) meat of the turkey. Goose is generally considered exceedingly fatty and thus rather too rich and indigestible for modern stomachs. However, this opinion is only true of the commercially produced Christmas Goose, fattened up on potatoes and other starchy foods to reach more profitable fat-filled weights. Like all water birds, the goose goes heavily to fat if over-fed. But needless to say, if you are keeping your own birds, it is not necessary to over-feed them. If you prefer a slightly smaller, much less fatty Christmas bird, choose your breed and feed accordingly. Then you will appreciate why the French, in their culinary wisdom, have stuck to goose for Christmas dinner. By comparison, turkey tastes like overgrown, dried-out broiler meat.

Geese are very bad-tempered. In the breeding season, the male is positively dangerous if you get too near the nest. Geese are also very noisy, with a loud jarring call which can be frightening. They have traditionally been kept, since Roman times, to guard against intruders by sounding the alarm. Their trouble is that, as watchdogs, they are unreliable, tending to give the alarm far more often than is justified. Nevertheless, geese can be very effective for keeping intruders out of orchards at picking time, and this is often the main justification for keeping them.

BREEDS

The two common white breeds of goose are the Embden and the Roman. Both are fine foragers. The Embden is considered to be a rapid grower and excellent for producing a less fatty Christmas bird.

For a fattier Christmas bird, if that is what you prefer, get hold of a Toulouse, a grey and white goose breed. Chinese geese are very decorative, but their meat is highly flavoured and they are poor foragers. Muscovies are the easiest of all to keep, being hardy and excellent foragers. They are also very useful as foster mothers for raising other breeds of geese. The meat of the Muscovies is rather darker in colour than we are used to, but very tasty if killed young (drakes reach 7 lb at 3 months). Muscovies are useless as alarm sounders.

Since the keeping of geese has reverted to a fairly informal non-commercial affair, quite a large proportion of the geese now kept are in fact cross-bred. Most of these are standard white geese, however, with attributes near to the Embden.

BASIC NEEDS

Geese are grazers and do best if given plenty of room on grass-land - at under 20 to the acre. They do not take very happily to close confinement, though it can be done if necessary. Because of the noise, they are totally unsuitable for keeping in town gardens.

Geese will sleep out all year if necessary, but it is advisable to protect them from foxes at night. All that is needed is a dry shelter. No perches or nest boxes are required.

FEEDING

Geese were traditionally kept on commons, given little or no food beyond the grass, weeds and other scraps they found for themselves. If you have enough grassland, you can do the same, giving them a small feed of grain in water at night to get them into into the fox-proof house. For fattening, use a starchy mash of potatoes, scraps, grain meals and the like.

REARING

The usual arrangement is to keep one gander and two stock fe-males. These should hatch and rear 10 - 20 goslings every year, to be slaughtered in early winter. But geese are choosy who they mate with and you must allow some time for new partners to get acquainted.

Alternatively, a broody hen will hatch and rear up to 6 gos-lings. But the eggs are too big for her to turn, so mark them and turn them every day. They also need sprinkling with water like duck eggs. Goose eggs hatch in 28 - 35 days. Goslings brought up by a broody hen tend to wander less far.

KILLING

Geese are big birds to kill. The method recommended is to stun the bird first with a sharp blow to the back of the head. This saves any further pain. Then find the point at the back of the neck where the neck joins the head. Insert a knife into the gap to sever the artery and pierce the brain. Pluck immediately, in spite of all the blood. Geese are not easy birds to pluck.

SOURCE OF MORE DETAILED INFORMATION

"Ducks and Geese", published by HMSO (see Chapter Eleven).

SOURCES OF STOCK

I do not know of any commercial sources of stock geese. You will have to track down a local person who keeps geese and persuade him to let you have a sitting of eggs to put under a broody in early Summer.

13 turkeys

WHY KEEP TURKEYS?

There is only one good reason for keeping turkeys: because you like the meat. They are more recently domesticated than fowls and are thus more troublesome. When young, they need considerable care to protect them from wet and cold. They are also prone to disease and parasites.

If I have not already made it clear, I should now state that turkeys have never appealed to me. I find the meat inferior and the bird both ugly and bothersome. Nevertheless, I will try to put down the basics fairly in this chapter as in the previous two.

BREEDS

Hybridisation has now hit turkeys in a big way. They grow bigger and faster than ever before. I do not know of any sources of the older pure breeds such as the Cambridge Bronze, or the Norfolk Black.

BASIC NEEDS

Like the fowl, and unlike ducks or geese, 'turkeys in the wild are
forest roamers. But they come from the warm climate of Central
America and have not been in Britain long enough to have adapted
fully to our wet, mild climate. The primary needs of the turkey
are shelter from wet when young; plenty of fresh air and ample
exercise. Since the first of these tends to conflict with the other
two primary needs, turkeys have a reputation for being rather
tricky to keep.

Adult turkeys are hardy enough. They are happy to roost in
trees all through the Winter. However, young stock must be very
carefully protected from anything wetter than a light Summer
shower.

Turkeys are very prone to a nasty disease called blackhead.
Because this is spread by fowl, turkeys must never be kept together with fowl.

SYSTEMS

Beginners are not recommended to attempt hatching and rearing
up young turkey chicks, not until they have got the hang of things.
If you want a few turkeys for killing at Christmas, buy them from
a turkey farm at about six weeks old in August. They will be expensive (£1.75 each this year), but will grow to a Christmas live
weight of 15 lb for hens and 25 lb for stags. That way, you avoid
the worst problems of rearing. However, your turkey 'poults'
(chicks) will almost certainly have been kept inside and fed on
mash with anti-blackhead additives. You will need to wean them
off all that gently. Keep them on turkey rearers' mash inside
in a good, dry, well-ventilated house until they 'shoot the red'.
In other words, they show a definite red colour about their neck
caruncles, those strange protrusions in front of their necks.
Once your poults have 'shot the red', they are pretty hardy. They
can be let outside and changed over to a more natural diet.

Young turkeys can be kept in deep litter houses, straw yards
or on free range. They will roost in trees if encouraged when
first let out. Alternatively, provide them with a fox-proof night
shelter with perches and one side covered only in mesh or netting.
That will give them the fresh air they need.

FEEDING

Turkeys can be fed as fowls, but they prefer feeding little and often, so three feeds per day are better than the normal two. To fatten them, use oats or barley. Feed plenty of grit. Green food is as essential as for fowl. They are better grazers than fowl, being particularly partial to young grass, clover and dandelions.

REARING

If you set yourselves up with a breeding pair, the hen can be left to do her own hatching and rearing, or a broody chicken will do the job. Mark the eggs and give her help with turning if she finds them too heavy. Turkeys reared by hens have the advantage that they do not wander so far afield.

KILLING

Young turkeys ready for table are killed and plucked in the same way as fowls.

DISEASES AND PARASITES

The big danger is blackhead. Look out for lethargy, loss of appetite and yellow diarrhoea. In some cases, the head darkens or goes black. Death follows soon. Drug treatments (eg. Entramin A) are available. Turkeys are also prone to colds and to plain diarrhoea (see Chapter Ten). Lice are also a particular problem in turkeys. Dust with pyrethrum powder to clear the birds. Turkeys also sometimes get large ticks on their heads, which are best burnt off carefully.

SOURCES OF MORE DETAILED INFORMATION

"Turkey Farming", by R.Feltham, published by Faber.

14 bantams, guinea fowl, peafowl and decorative waterfowl

This is the 'odds and ends' chapter. You may well start your interest in poultry from an economic, practical point of view. But there is no doubt that a lot of people go on from this after a few years to expand their poultry-keeping into a hobby for pleasure. One of the pleasures is trying your hand at the more obscure types of poultry, kept as much for their decorative value as for their produce.

BANTAMS

Bantams are miniaturised fowl (a bit like toy poodles are to real ones). They usually come at about half full size. All the common pure breeds have been bantamised: the Rhode Island Red, Light Sussex, Leghorn, etc. In addition, a considerable number of other bantam breeds have been developed more for their decorative interest than as miniature laying hens. Among these are:
> the Old English and Indian Game Bantams
> the Silkies, Frizzles and Rumpless Bantams
> the Pekin, Nankin and Japanese Bantams
> the Cochin and Brahma Bantams - and so on.

Earlier this century, the keeping of bantams was very popular. It was argued that, since they ate half the food and needed half the space, they were a better proposition for small gardens. Not only that, but there was the added interest of showing birds at local, regional and national poultry shows and competing for prizes, rather like at dog shows. That was alright in the days when 100 eggs per year was acceptable, but the laying performances of the bantam breeds since then have fallen way behind those of the modern hybrids. If it is eggs you want, then bantams can no longer compete, even in a small garden.

But if you are attracted by the beautiful plumage and unusual shapes of the bantam breeds, then keeping bantams is a hobby you might consider. Most of the big regional agricultural shows still have sections for show poultry and in the North of England, bantams can also be seen at specialised local shows. I suggest that as a first step, you visit one or two shows, and at the same time try to make contact with local bantam fanciers, who, like most minority enthusiasts are generally very helpful to beginners.

Be warned in advance, however, that bantam cocks have a very high-pitched and generally pretty loud crow. If you are planning to keep them in a small garden in a built-up area, this is a big disadvantage. Bantams can also be very nervous and flighty (like toy poodles) and need high fencing and regular wing clipping to keep them in.

Expect about 100 to 150 eggs per year weighing 1 to 2 ounces. Feed them as hens, but halve the quantities. Bantams are generally excellent sitters and foster-mothers, and a single bantam hen of a game breed can pay for her keep on that quality alone.

My personal attitude to bantams, for what it is worth, was conditioned by my ex-neighbour's bantam cock who used to lead his flock every evening to roost in a tree only twenty yards from our bedroom window. We were woken regularly at four in the morning by a most dreadful high-pitched raucous yell, the like of which I never want to hear again. I can appreciate the fine plumage of bantams and enjoy viewing them over at shows, but they are such twitchy, ungraceful creatures that I doubt I shall ever actually keep any.

GUINEA FOWL

If you have never seen Guinea Fowl, they look like small grey turkeys without tails. They are not a very domesticated form of poultry but yield tasty game-like meat. They are also the best insect eaters of all poultry and can be used to keep potatoes clear of beetles (at one bird per acre).

They need no housing, much preferring roosting in trees. Even given nest boxes, they hide their eggs all over the place. Confinement does not seem to suit them too well, although some people keep them inside fences with their wings clipped.

In the wild, the Guinea Fowl is monogamous, but in captivity the male can be persuaded to take two wives (is there a moral in this?). They can either be left to rear their own young or the eggs can be put under a broody hen in the usual way. They hatch in 26 to 28 days. Keep the chicks dry until they are well feathered and give them plenty of protein or access to lots of insects. Eaten at about two years of age, the birds reach about 4 lb.

Guinea fowl have a very loud high-pitched call. The books say they can be used like geese, as alarm sounders, but I have two sets of friends who have tried this and found them total failures. One lot never shut up, and the other got eaten by a fox before giving the alarm...

PEACOCKS

Peafowl used to be kept for their meat, which tastes rather like pheasant. Now they are of interest purely for the male's splendid tail-feather display. They do indeed look splendid on the lawn, but remember that they are capable of considerable crop damage and should therefore be wing-clipped and confined.

The shrill cry of the male during the mating season is very noticeable and not altogether pleasant. The peahen is a poor sitter and very easily disturbed off the eggs. They are prone to coccidiosis.

If you want a pair, look for small-ads in Country Life and the Country Gentleman's Association Magazine. Keep the chicks dry, feed them near a suitable barn or outhouse which they will then make their permanent home.

OTHER POULTRY

Swans are graceful birds to have around the place, if you have water on your property. They are wild and powerful, not suited to domestication. The best you can do is feed them regularly on the water, provide them with nesting material near the water's edge at the right season and just enjoy having them around.

There are, in addition, many varieties of ornamental water fowl, such as the ducks, moorhens and wild geese seen on the lakes of our parks. The needs of such birds are modest, beyond a suitable lake or river, but breeding can be tricky and the assistance yet again of the good old broody hen can be useful.

A fine selection of water fowl is available from John Hall, Red House Farm, Chediston, Halesworth, Suffolk. His latest catalogue includes the following:

> 4 varieties of swan;
> 34 varieties of goose, from the Hawaiian at £300 per pair to the Egyptian at £10 per pair;
> Over 80 varieties of duck, from the hand-reared Bufflehead at £250 per pair to the Dutch Call Duck at £5;
> Bantams, red jungle fowl, peafowl and performing pigeons (?).

APPENDIX 1: CHECK-LIST OF TASKS TO COMPLETE BEFORE
ORDERING FIRST STOCK

	Chapter (for details)
* Check no anti-poultry regulations in your area	1
* Are you and your family agreed on who does what work and what happens when you are away?	1
* Decide on your System of Management	3
* Decide what type of stock you will buy, and how many	4
* Decide what you will be feeding	5
* Stock up with adequate quantities of feed and nesting material	5
* Obtain a drinker plus, depending on your system, receptacles for grit, oyster shell, main feed and protein	5
* Obtain and/or build adequate housing and fencing both for rearing and for adult stock	6

APPENDIX II: CHECK-LIST OF SUGGESTED REGULAR ROUTINES

	Chapter (for details)
Daily	
Feed morning and afternoon	5
Collect eggs three times, clean and store them	9
Every 2 to 4 days or twice weekly	
Check water in drinker and supply of grit and shell if given	5
Feed protein concentrate	5
Every fortnight	
Move the henhouse to a new spot	10
Clean out the house thoroughly and replace nesting straw	6
Every month	
Clean the drinker out thoroughly. Put new garlic in drinker	5 and 10
Every three months	
Check feeding quantities are right. Adjust if needed	5
Every early Spring	
Buy in new day-old stock (can be done every other spring)	8
Every Autumn	
Cull, or kill off old stock	8
Speed the moult	10

APPENDIX III: IDEAS LIST OF MORE INTERESTING WAYS TO EAT EGGS

Fast ways with eggs (under 10 minutes)

Soft boiled	5 minutes in boiling water - cool fast
Hard boiled	10 minutes in boiling water
Poached	add a drop of vinegar to water - boil 1 minute
Fried	either spoon fat on or cook under lid
Scrambled	beat eggs well - butter pan - stir when sticking to pan - serve creamy
Oeufs sur le plat	butter a fire-proof dish - break egg in - heat till white sets
Omelette	beat eggs - have butter just smoking - shake to stop sticking - serve creamy in centre - fillings endless
French toast	soak bread slices in beaten egg/milk mix - fry till golden
Swiss eggs	$\frac{1}{2}$" of milk in baking dish, then eggs broken in, then grated cheese - bake for 10 minutes

Uses for hard-boiled eggs

Curried eggs	mix yolks with curry powder - return to white halves
Egg cutlets	add chopped to thick white sauce - cover cutlets with breadcrumbs and fry
Pickled eggs	pour spiced vinegar boiling over eggs - when cool, seal - use after 14 days
Creamed eggs	pour spoonful thick seasoned white sauce on buttered toast - lay on strips of white covered with sprinkled yolks

| Portuguese eggs | lays sliced eggs in baking dish - cover with fried onion/tomato mix, then bread crumbs, grated cheese and butter - bake for 10 minutes |
| Stuffed eggs | like curried eggs, but with mayonnaise, chopped onions etc. mixed in - try other mixtures |

Eggs broken onto various delicious pre-cooked mixtures, to poach on the mixture

| Chatchouka (N. Africa) on fried pepper/tomato mix |
| Oeufs en Matelote | on herbed, seasoned red wine, later thickened and poured over eggs |
| Egg bouillabaisse | on onion/garlic/tomato puree/stock mix, served as soup |

Ways of serving poached eggs

With mushrooms	make puree of mushrooms with flour, butter and cream - pour over eggs
With tomato	do same as above with freshly made puree of tomatoes
Eggs Mornay	with cheese sauce on puree of potatoes
Eggs Mikado	served cold on a bed of rice/celery/peppers salad
Eggs in aspic	add cooling aspic and shredded ham over eggs

Using up Summer over-production (consult detailed recipes)

Hot Souffle	yolk/butter/flour mix fluffed up with whisked whites - flavoured and cooked
Creme caramel	egg/milk/sugar mix cooked gently on a caramel base
Confectioner's custard	
	no backyarder should buy custard powder
Quiche	a pastry flan filled with egg mixed with bacon and cheese, onion, tuna, corned beef and cooked until firm
Lemon curd	a thick custard of eggs, lemon juice and lemon zest

Egg based Sauces

| 'Cheezle' | 3 oz grated cheese to 1 beaten egg - heated and stirred continuously until |

	thick (invented by an ex-flatmate of mine, fast but tricky and good on vegetables)
Mayonnaise	egg yolks with olive oil added gradually, stirring vigourously
Hollandaise	egg yolk/ spiced vinegar mix with butter added gradually over gentle heat

Thickening a cooking sauce with Eggs and Cream

Mix the eggs and cream and add hot sauce gradually to this mix. Reheat but do not boil.

Drinks

| Zabaglione | yolks, sugar and Marsala wine whisked up over a gentle heat and served warm |
| Advocaat | similar way, but with brandy and almond essence. Served cold. |

APPENDIX IV: UK AND EEC EGG GRADING SYSTEMS

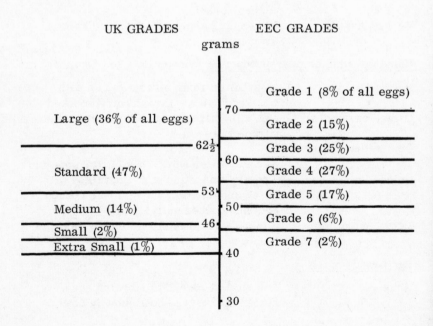

LIST OF REFERENCES

GENERAL GUIDES, GEARED TO PRE-BATTERY COMMERCIAL
POULTRY KEEPING

There must have been literally scores of titles on this subject
written between 1890 and 1960. I cannot possibly cover them
all. If you are interested, see what you can pick up in your
local second hand bookshops. Here are a few that I have come
accross:

W.Powell-Owen. Fortunes from eggs. (circa 1920)
S.H.Lewer. Wright's book of Poultry. Waverley Books (1918)
F.E.Wilson. Poultry Keeping and how to make it pay. Pearson
        (circa 1920)
A.Thompson. The Complete Poultryman. Faber (1952)
A.Thompson. Keeping chickens for profit. Nicholson and
        Watson (1952)
L. Robinson. Modern Poultry Husbandry. Crosby Lockwood
        (1960)

.... I could go on and on. From most of them the backyarder
can learn a great deal.

GENERAL GUIDES, GEARED TO BACKYARDERS

I have found virtually no books in this category, which is one of
the reasons why I wrote this book, but the nearest is:

J.Worthington. Natural Poultry Keeping. Crosby Lockwood
        (1964/1970)
- a very personal 'how I did it' book by a true pioneer. There
is a lot to learn here, but you have to really look for it.

Although there is a shortage of books of this type, there are a
few booklets designed to help the backyarder:

Keeping Chickens. FAO Better Farming Series (1970)
- a guide for African peasant farmers

<u>Poultry Keeping for Housekeepers</u>. Ministry of Agriculture
Growmore Leaflet No.5 (1939)
- part of the War effort with lots of the 'official line' on boiling
scraps etc.
<u>Starting in Poultry</u>. South Australian Department of Agriculture
Bulletin No.458 (1960)
<u>Domestic Poultry Keeping</u>. Ministry of Agriculture (1975)
- recently produced to meet the demand for advice; very super-
ficial, with the emphasis on intensive systems and mixed mash
feeding.
D.Bland. <u>Poultry for the Garden</u>. published privately (1975)
- a nicely produced 40p booklet. Good coverage but again not
'natural' enough in its approach for my liking.
<u>Poultry Keeping for Beginners</u>. Ditchfield's Little Wonder Books.
- dated and not very well done.
<u>Poultry Keeping</u>. Young Farmers Club Booklet No.5 (1945).
- thorough and well illustrated, but out of print, I believe.

BOOKS FOR LEARNING ABOUT THE MODERN "POULTRY
INDUSTRY"

Card and Nesheim. <u>Poultry Production</u>. Lea and Febiger, USA.
(1972)
- a very high standard American commercial manual.
D.O.Griffeths. <u>The Poultry Industry in England and Wales</u>.
article in the Royal Agricultural Society of England
Journal, Vol 134 (1973).
<u>Poultry World</u> - <u>the</u> trade magazine.

GENERAL "BACKYARD FARMING" GUIDES, WITH SECTIONS
ON POULTRY

C and E.Robinson. <u>The Have-More Plan</u>. Macmillan, USA (1947)
and more recently Gardenway Publishing, USA.
- From the original pre-war American back-to-the-land movement. Good reading to inspire you with confidence.
William Cobbett. <u>Cottage Economy</u>. reprinted Chivers (1966)
and Landsmans Bookshop (1975)
- The original backyarder. A joy to read.
John Seymour. <u>Self-Sufficiency</u>. Faber (1973).
- a modern Cobbett, equally readable.

There are others. None of them very good on poultry keeping.

This branch of backyarding tends to be treated as all-too-easy common sense stuff, compared to the complexity of smoking bacon and wheat husbandry. One general small-scale farming book, however, is detailed on poultry:

John Hayhurst (ed). The Smallholder Encyclopaedia. Pearson
(1950)
- a fine work of reference, but arranged alphabetically, which has always seemed pointless to me for a specialist work. Out of print.

MORE ON THE PHILOSOPHIES OF NATURAL FARMING

G. Henderson. The Farming Ladder. Faber (1944)
- a great farmer philosopher.
Any book by Newman Turner, another fine farming writer. Obtain the latest booklist from the Soil Association. (Walnut Tree Manor, Haughley, Stowmarket, Suffolk) and take your pick. They promote organic farming.

MORE ON DEEP LITTER

R.Feltwell. Deep Litter for Egg Production. Faber (1954)
Environmental Control in Poultry Production. Oliver and Boyd
(1967).
Lighting for Egg Production. Ministry of Agriculture Leaflet
No. AL 540 (1974).

MORE ON THE FOWL'S BEHAVIOUR PATTERNS AND BASIC NEEDS

D.Wood-Gush. The Behaviour of the Domestic Fowl. Heinemann
(1971).
- details of all the research relevant to the subject.

NATURAL VETERINARY MEDICINE

J. de Bairacli Levy. Herbal Handbook for Farm and Stable.
Faber (1963).
- This book stands on its own on the subject.

UNDERSTANDING HYBRIDISATION

Hybrid Chickens. Ministry of Agriculture Bulletin 180 (1959)

- very objective. An excellent introduction.

IMPROVING YOUR TRUSSING TECHNIQUE

Use the Ministry's picture-by-picture guide leaflet: Ministry Leaflet No. AL 428 (1975).

MORE ON COMFREY

L.D.Hills. Comfrey Report. Henry Doubleday Research Association, (1974). Convent Lane, Bocking, Braintree, Essex.
L.D.Hills. Comfrey. Faber (1976).

BACKYARD DAIRY BOOK
Andrew Singer and Len Street

This book is written as propoganda and its aim is to provide enough basic information to encourage readers to begin home dairy production. Thousands have reduced their dependence upon factory food by growing their own. or keeping chickens. Backyard Dairying is a further step towards self-sufficiency.

Contents

"The chapters dealing with goat management are most convincing"—*Undercurrents*

"A very useful book"—*Sunday Times*

"It is remarkably clearly written and will be welcomed by many questing clients in a practitioner's waiting room"—*Veterinary Record*

Copiously illustrated
88 pages 8½" x 5½"
ISBN 0 904727 05 X Hardback
ISBN 0 904727 06 8 Paperback

BACKYARD PIG & SHEEP BOOK
Ann Williams

BY ANN WILLIAMS

Ann Williams is a farmer and journalist, and for ten years worked as an adviser to the Ministry of Agriculture. Her book follows the highly successful approach of the "Backyard Series" and provides a thorough initiation into the delights of pig and sheep husbandry.

Contents

"For those who have already started, nothing could be better than this comprehensive guide" —*Undercurrents*

Numerous illustrations
185 pages 8½" x 5½"
ISBN 0 904727 45 9 Hardback
ISBN 0 904727 46 7 Paperback

BACKYARD BEEKEEPING
Bill Scott

Illustrated by Keith Spurgin

A basic introduction to the gentle art which removes much of the mystique surrounding it. Part one explains the life of the hive and bee, part two describes the equipment and materials, and includes a design for a D-I-Y hive. Part three covers potential problems and their solution. The final section looks at the harvest and what can be done with it. Bill Scott is author of our very successful book, Food for Thought, and runs a wholefood shop in Truro, Cornwall.

"To be a successful beekeeper with a happy colony and plentiful harvest, you will have to invest time, effort and knowledge, and that's what this book can give you even if you are a complete novice".—*Here's Health*

"Unhesitatingly, I would say this book is a *must* for anyone contemplating keeping a hive or two of bees. Its simple clear style inspires confidence right from the start. The basics of beekeeping are explained in the clearest possible way and it even includes a section on making your own equipment."—*The Soil Association News*

112 pages 8½" x 5½"
ISBN 0 904727 43 2 Hardback
ISBN 0 904727 44 0 Paperback

BACKYARD RABBIT FARMING
Ann Williams

Ann Williams, the author of the "Backyard Pig and Sheep Book" draws on her long experience of rabbit rearing and demonstrates how this activity can be made both profitable and fascinating. Apart from chapters on housing, feeding, management, breeding, diseases and harvesting there is, unusually, a long and detailed section on what can be done with pelts and fur in the way of curing, dyeing, weaving, spinning etc.

Contents
1. Why keep rabbits?
2. Rabbits past and present.
3. About the rabbit.
4. Making a start.
5. Housing.
6. Feeding.
7. Growing crops
8. Breeding.
9. Health.
10. Wild rabbits.
11. Meat.
12. Recipes.
13. Fur production.
14. Showing rabbits.
15. Angoras.

Bibliography

Illustrated
About 120 pages 8½" x 5½"
ISBN 0 904727 55 6 Hardback
ISBN 0 904727 56 4 Paperback

PRACTICAL SOLAR HEATING

Kevin McCartney

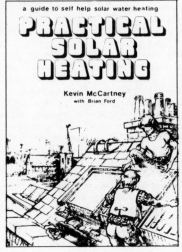

a guide to self help solar water heating

PRACTICAL SOLAR HEATING

Kevin McCartney
with Brian Ford

1. Solar Energy—what is it, why we should use it and how.

2. Basic Principles—Absorption, heat loss, greenhouse effect, heat and temperature, heat capacity and reaction time, tilt and orientation angles.

3. Solar Collectors—Function, types, surface finishes, materials, insulation casing and glazing.

4. Storage Tanks—Hot water cylinders, galvanised iron and plastic tanks, heat exchangers, cold feed tanks, expansion tanks, pressurised tanks and insulation.

5. Circulation—Thermosyphoning (gravity) and forced (pumped) circulation, pump and pipe sizes and materials, pipe lagging and controls.

6. Mounting the Collectors—Location, building permission, fixing over existing roof, removing roof and fixing collectors under glazing bars, wall collectors and free-standing collectors.

7. Choosing a System—Step-by-step guide to variations in methods of connecting collectors to storage tanks, frost protection and temperature boosters.

8. Plumbing for Solar Systems—Plumbing without a blowtorch, compression fittings, pipe types. With a blowtorch, capilliary fittings, low-cost plumbing.

9. Swimming Pools—Types of collector required, size, location and mounting, insulating the pool, effectiveness.

10. Examples—Commercial, Council and D.I.Y. installations.

11. D.I.Y. Collectors—Detailed plans for two types.

12. Installation Guide—Step-by-step instructions.

13. Survey of Manufactured Collectors.

Even in a climate such as ours up to 50% of our domestic water heating could be supplied by the sun. So far only cost has postponed the widespread use of solar energy in homes throughout the country. Now, however, the rocketing price of conventional energy sources has made solar water heating economically competitive. Furthermore new plumbing materials and techniques now make do-it-yourself installation quite feasible, thereby greatly reducing the capital outlay.

SMALL SCALE WATER POWER
Dermot McGuigan

Do you like the sound of a rushing brook? It sounds even better when all that energy is harnessed and is lighting and heating your home, or providing the power for a farm or small industry.

If you live near a brook or a river, or are thinking of building near one, this book is a must for you. It tells you how to best tap that source of power to meet your electrical needs.

With the age of cheap energy drawing to a close, interest in water power is growing fast. Those sharing that interest will find this is an excellent sourcebook, telling how to estimate the power in a stream, where the most suitable equipment can be obtained, and at what price.

There are detailed descriptions of working installations, with an analysis of their costs. The recent innovations that have cut the costs of hydropower equipment are explained.

You'll also find here: Information on dams, fish passes, spillways, pipelines, drivers, alternators, governors and legal aspects—plus a detailed manufacturer's index.

SMALL SCALE WIND POWER
Dermot McGuigan

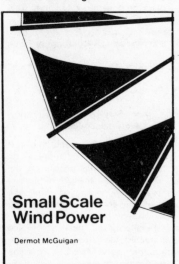

Wind power, harnessed centuries ago by man, is fast gaining new popularity.

Is it for you? This book lets you decide. It presents a survey of numerous working windplants, telling the many purposes which wind power is best suited to serve.

There's an appeal to wind power. It's there— we hear it every day. It's non-polluting and inexhaustible. And it's free, after the initial installation of the windplant.

Homeowners, farmers, small industries are looking again at wind power. Join them. Dermot McGuigan makes it easy to take an informed look. You can reach a decision based on scientific fact, not hopes or promises.

Here, in one book, are all of the pluses and minuses of wind power, and the many types of wind machines now on the market, written so that you can understand them. If you are interested in alternative energy, you'll find this book engrossing. If you want to reduce that growing fuel bill, this book may show you how.

BE NICE TO NATURE
Natural Pest Control
Greet Buchner & Fieke Hoogvelt

Contents

How can we protect our roses from greenfly, chase stray dogs from the garden, insects from the bedroom and mice from the kitchen. With aerosols and chemicals? Not if we care about the environment. However if we abandon chemicals will our plants be devoured and homes overrun? Bio-dynamic gardeners, Buchner and Hoogvelt, don't think so and have compiled some gentler, organically-based alternatives.

"The author's treatment for aphids is to spray with nettle manure. It's certainly much cheaper than a proprietary spray"—*Farmers Weekly*

Illustrated with line drawings
90 pages 8½" x 5½"
ISBN 0 904727 07 6 Hardback